The Institute of Biology's
Studies in Biology No. 55

Arthropod Vectors of Disease

James R. Busvine
D.Sc., F.I.Biol.

Professor of Entomology as Applied to Hygiene
London School of Hygiene and Tropical Medicine

Edward Arnold

First published 1975
by Edward Arnold (Publishers) Limited
25 Hill Street, London W1X 8LL

Board edition ISBN: 0 7131 2500 4
Paper edition ISBN: 0 7131 2501 2

Printed in Great Britain by
The Camelot Press Ltd, Southampton

General Preface to the Series

It is no longer possible for one textbook to cover the whole field of Biology and to remain sufficiently up to date. At the same time teachers and students at school, college or university need to keep abreast of recent trends and know where the most significant developments are taking place.

To meet the need for this progressive approach the Institute of Biology has for some years sponsored this series of booklets dealing with subjects specially selected by a panel of editors. The enthusiastic acceptance of the series by teachers and students at school, college and university shows the usefulness of the books in providing a clear and up-to-date coverage of topics, particularly in areas of research and changing views.

Among features of the series are the attention given to methods, the inclusion of a selected list of books for further reading and, wherever possible, suggestions for practical work.

Readers' comments will be welcomed by the author or the Education Officer of the Institute.

1974

The Institute of Biology
41 Queens Gate,
London,
SW7 5HU

Preface

Paradoxically, in the last moribund phase of colonialism, the developed countries began more and more to work for the amelioration of the lives of tropical peoples. This struggle is organized either collectively through the World Health Organization, the Food and Agriculture Organization, etc., or individually through various national agencies such as the Overseas Development Administration in Britain. To implement the programmes of these bodies there is a need for people with specialized training to opt for tropical service.

It is to be hoped that this book may fire the interest of some young people with the fascination of that strange life form, the arthropods, the diseases they transmit and their effects on men of many races.

London, 1975

J. R. B.

Contents

1 Introduction

A very substantial proportion of the most serious human infections are transmitted by arthropods. Most of the more serious tropical diseases are involved, for example, malaria, yellow fever, sleeping sickness, plague, typhus, relapsing fevers, various kinds of filariasis and encephalitis, Chagas' disease, Rocky Mountain fever and scrub typhus. To understand the scale of some of these afflictions it should be realized that about 1,900 million people live in potentially malarious regions. Due to human efforts, the risk has been circumscribed and now only about 400 million live in areas of high endemicity. Efforts to reduce this still further have, however, been unsuccessful. As a major scourge malaria is being challenged by filariasis which is currently estimated to be responsible for over 250 million cases at a given time, with the number increasing daily. Many of these may end with the horrible disfigurement of elephantiasis. Onchocerciasis, an analogous infection due to nematode worms in the tissue, affects about 20 million people, mainly in West Africa. About one in ten of them will eventually become blind. Chagas' disease, in Central and Southern America, affects some 10 million rural dwellers, leaving incurable heart damage from which many will die in middle life.

In the past, the significance of these diseases for historians (and through them, for ordinary people) has depended on the impact which they made on civilized communities. Thus, few of us have not heard of plague and its grisly synonym, the Black Death. Typhus, though equally grave, tended to be the concomitant of war, starvation and various camp diseases, so that it did not stand out so sharply. Malaria was represented by the relatively minor ague of Northern Europe, though known as a killer in the south. Violent, lethal epidemics, especially in Europe, were bound to have made more impression than the decimating diseases of the tropics which, until the nineteenth century, were largely mysterious and unknown. When the exploration and colonization began in earnest, it was (not unnaturally) diseases of the white adventurers which attracted attention. Malaria gave West Africa its reputation as the 'White Man's Grave'; the construction of the Panama Canal called attention to the ravages of yellow fever; African explorers found their horses and cattle dying from nagana. Towards the end of the nineteenth century, there began an intense effort to discover the ways in which these diseases spread. In 1878, Manson, working in China, discovered a human parasitic nematode in the bodies of mosquitoes, though he did not then realize that transmission was via the bite. In 1895 Bruce proved the carriage of nagana by tsetse flies and eight years later with Nabarro, showed a similar cycle with sleeping sickness. Meanwhile, Ross had demonstrated mosquito transmission of malaria, further elucidated by Grassi, Bignami and Bastianelli in Italy. About the same time Simond described the first rat transmission of plague. A year or two later in Cuba, Reed, Carrol, Lazear and Agramonte proved the transmission of yellow fever by a mosquito. In 1906 Ricketts in Montana demonstrated transmission of Rocky Mountain fever by ticks. Louse-borne diseases followed: relapsing fever in 1907 by Mackie in India and epidemic typhus by Nicolle and others in Tunis.

Most of these early pioneers were physicians, several of whom fell victim to the diseases under study and succumbed. Later, other professions were involved, in the search for remedies by immunization, drugs or vector control. In the third and fourth decades of this century discoveries of safe and reliable prophylactic drugs and effective means of immunization against disease began to make the tropics very safe for expatriates from the temperate zone and the more advanced inhabitants. The main sufferers now are the peasants in rural hinterlands as well as large numbers dwelling in slum concentrations round many tropical cities. For various reasons, prophylactics have seldom been feasible for their protection and this has only become possible with the development of modern synthetic insecticides for control of the disease vectors.

The early and impressive success of DDT suggested that the problem was largely solved and encouraged somewhat over-optimistic hopes of the total eradication of such immense and long entrenched scourges as malaria. Subsequently, the snags began to emerge, with setbacks such as the insecticide-resistant strains of many vectors and the anxiety about environmental pollution. Despite these difficulties, much progress has been made. The indications are that further advances will demand more detailed knowledge of the arthropod vectors which are the subject of this small book.

2 The Pathogens

The micro-organisms which cause disease are a very heterogeneous collection, beginning with extremely small viruses. These start at an order of magnitude comparable to a large protein molecule, say 10 nm, with the largest about twenty-five times greater.* They are necessarily rather simple forms of life, and all are parasitic in cells of larger organisms and cannot be cultured in non-living media. Somewhat larger are the *rickettsiae* extending over the range from large viruses to very small bacteria, which may be rather less than 1 μm in diameter.* The rickettsiae are also exclusively parasitic, always involving arthropods at some stage, but also extending to vertebrates. At one time they were thought to be intermediate between bacteria and viruses, but they are now considered to be specialized (? degenerate) forms of the former. They are, for example, susceptible to antibiotics, whereas viruses are not. Next to be considered are bacteria, a large and heterogeneous assembly, with free-living as well as parasitic forms. From these are met two representatives: the plague bacilli (about 1.8 × 0.6 μm) and the *spirochaetes* of relapsing fevers, (8 to 15 μm long by 0.3 μm thick). The former belongs to a group with thick rigid walls; whereas the spirochaete wall is thin and flexible and the spiral cell intertwined with a long slim filament used for locomotion. *Protozoa* are generally only slightly larger than bacteria, but they represent a higher form of cell life with a true nucleus containing chromosomes; these separate by mitosis during cell division. Among the protozoa are encountered the malarial parasites (*Plasmodium* spp.) about 2–4 μm in diameter, also the larger flagellates, such as the *Leishmania* spp. of kala azar and oriental sore, and the long *Trypanosoma* of sleeping sickness, which may extend to 12 or even 40 μm long. This completes the unicellular micro-organisms, but it is convenient here to include certain parasitic nematode worms, the infective stages of which are about 300 μm by about 6 μm thick. Even so the largest of these pathogens must weigh only about 10^{-5} g; this compares with 10^{-2} g as an approximation for most of the insect vectors and 10^{5} g, the weight of a large man.

In considering the ecology of micro-organisms it is realized that they are, to a large extent, 'in passive slavery to forces of the environment' (Julian Huxley's words). Flagella or cilia may help them to move voluntarily a metre or so in an aqueous environment, but in general they are dependent on the chance forces of wind and water movements.

* A millimicron (nm) is one-millionth of a millimetre. A micron (μ) is one-thousandth of a millimetre.

This is not so much a serious limitation for organisms which thrive saprophytically in soil or some such extensive habitat. But for the small fraction of them which have adopted the habit of living parasitically on animals or plants, it is a serious matter. Their habitats are of limited size, and by microbial standards usually very remote from one another. Furthermore these breeding sites are mortal, so that a transfer to a new one is eventually essential. If the parasite causes disease, a move may be necessary rather soon.

There are various solutions to this problem. If the host is an animal, the micro-organisms can colonize a new one during direct bodily contacts which their hosts make in social, familial or sexual intercourse. Alternatively the micro-organisms can stay in, or invade, the gut and the spores can be emitted with the faeces, with the possibility of contaminating food or a drinking source. If the host is a vertebrate animal, they can be coughed or sneezed out in droplets of mucus, to float in the air and perhaps be inhaled by another host.

In the course of evolution, a remarkable series of adaptations took place to involve alternation between two different sized organisms: a large animal (or a plant) and a small creature such as an arthropod (commonly an insect). It is the vertebrate–arthropod cycle that is the concern of this book. This alternation was found to be beneficial to micro-organisms which had originally developed as parasites of arthropods as well as those which had begun as parasites of vertebrates. For arthropod parasites, the vertebrate alternative represented a very large and long-lived reservoir, capable of reinfesting a substantial number of new arthropods over a long period. The parasites originally in vertebrates, on the other hand, found a new long-range transmission method. The arthropods (particularly the insects) were especially useful, because many of them habitually associate with large animals, upon which they themselves are parasitic.

When human or animal diseases are concerned, it is normally considered that the arthropods which transmit them are the 'vectors' responsible for spreading the disease, and the vertebrates as the 'reservoirs' of infection. In some circumstances the roles are reversed. For example, a man infected by malaria or yellow fever, whisked into a new continent by aeroplane, could be responsible for spreading the disease alarmingly, though we would call him a 'carrier' perhaps rather than a vector. Again, some arthropods can act as long-lasting reservoirs, though less commonly than vertebrates, and on the whole, the terms are most conveniently used as mentioned earlier. Whatever the origin of these double parasitic cycles, involving the radically different hosts, it is certain that the extension from a single host arrangement must have presented considerable difficulties to be overcome by evolutionary adaptation. Some of these will now be considered.

2.1 Vector efficiency and specificity

(i) *Proliferation in the new host* must involve adaptations analogous to the initiation of the original parasitism. These include the development of enzyme systems capable of absorbing and metabolizing food sources in the new host's tissues, and also of combating the protective immunological reactions of this host. These are biochemical modifications of considerable complexity, and they are presumably responsible for the fact that a particular micro-organism will only thrive in a limited number of vectors and hosts. Thus, the malaria *Plasmodium* will develop in anopheline mosquitoes, but not in culicines, the sleeping sickness *Trypanosoma* will develop in tsetse flies but not in other kinds of biting fly, and so forth.

(ii) *Efficient transfer mechanisms* can be of a great many kinds, some very complex, and there is evidence of evolutionary trends towards the more efficient forms. In the simpler type there is a considerable fortuitous element and the vector is only an alternative to other methods of transmission, such as contagion or by water- or air-borne routes. These vectors may not be parasitic themselves. Thus, certain omnivorous insects which frequent human dwellings such as the housefly or certain cockroaches, are quite liable to visit human faeces for food or egg-laying (if these are accessible) and thereafter walk over and perhaps excrete on, human food. These habits obviously involve the possibility of transmission of enteric disease germs, though no specific disease relies exclusively on it. It is noteworthy that this kind of 'accidental' vector involvement is likely to be very handicapped by the short period over which the vector remains infected. There would clearly be survival value for a pathogen which developed a more permanent stay in the vector preferably by using it as a second host. This has happened in a variety of different ways, reaching over increasing degrees of complexity and efficiency.

A relatively simple system is exemplified by certain tapeworms, e.g. *Dipylidium caninum*, of the dog, the eggs of which are voided in the main host's faeces. These eggs may be swallowed by flea larvae and then the flea becomes parasitized. Dogs become reinfected when they swallow infected fleas and occasionally children who fondle their pets indiscriminately are infected. In this cycle, the infection of the flea vector is rather haphazard and it would seem easier to ensure infection when it took a meal. In the first place perhaps, this was fortuitous, due to contamination of the mouthparts. This method of transfer is, however, not very efficient. The small traces of blood on mouthparts might not include parasites and in any case, would soon dry up, which would harm many pathogens. Therefore, such mechanical transfer would only be likely to succeed when insects, disturbed during a feed, immediately resumed feeding on another animal. The next evolutionary step was, presumably, the proliferation of the parasite in the pharynx or gut of the vector. This is what happens in tsetse flies with the trypanosome responsible for 'nagana', a disease of imported cattle and horses in Africa. And something analogous seems to occur in the transmission of kala azar and oriental sore by the sandfly *Phlebotomus*. Chances of infection are, perhaps, magnified in plague fleas, in which the plague bacilli multiply in the pharynx, which then becomes blocked, so that the germs are regurgitated when the flea tries to take its next meal.

In other cases, the pathogen multiplies in the gut of an insect, without blocking it, and finally passes out with the insect faeces, which thus become infective. To enter another vertebrate host, these faeces must be scratched into an abrasion or enter a small wound. Alternatively, they can dry up to a fine powder, liable to float about in the air, and perhaps become inhaled, or else enter a delicate membrane of the vertebrate such as the conjunctiva of the eye. Louse-borne typhus and Chagas' disease (which is transmitted by large blood-sucking bugs) are two infections carried in this way. In a rather more complex cycle the micro-organisms are not restricted to the gut, but invade other tissues of the arthropod vector, which thus become a more thoroughly parasitized host. Obviously, this complicates the problem of reinfesting the alternate vertebrate host. In the simplest solution the parasite has no escape from the vector and can only reinvade its large host when the vector is swallowed and digested by the latter. An example is louse-borne relapsing fever, of which the causative spirochaetes are confined to the insect's body cavity. Infection can occur when primitive people burst them between finger nails or even between the teeth.

There are obvious drawbacks to a system where the vector has to be killed for the parasite to re-enter the vertebrate host, for one thing, it can only reinfect one such host, whereas, if reinfection is via the bite of an arthropod it could transfer the pathogen to several new hosts. A somewhat crude example of this is provided by filariases, parasitic worm diseases, carried by insects. In Bancroftian filariasis, carried by house-haunting mosquitoes and in onchocerciasis, carried by blackflies, the worms parasitize tissues of the insects. In the final stage, infective forms invade the proboscis and burst through the cuticle when the insect feeds. They are then able to enter the wound and reinfest man. In probably the most efficient mechanism which has been perfected by evolution, the parasites travel through the vector's body until they reach the salivary glands. In this way, they obtain easy entry to the blood system of a new vertebrate host, since saliva is normally injected in the process of taking a blood meal. This is the cycle underlying the transmission of quite a number of serious diseases, including malaria, yellow fever, sleeping sickness, dengue, mosquito-borne encephalitides and some forms of tick-borne relapsing fever.

It can be noted that in all cases where the cycle depends on the vector acquiring and transmitting infection by blood meals, it is most important whether the biting arthropod is catholic in tastes or whether it feeds regularly on the same species of host. Serious human disease vectors are those which feed commonly on man. Anopheline mosquitoes which never enter human dwellings to feed and prefer other vertebrates, are not vectors of malaria. On the other hand, tsetse flies which feed indiscriminately on wild game and on domesticated horses and cattle, may carry to the latter the fatal germ of nagana from the reservoir in the unharmed wild animals.

(iii) *Survival in the vector.* It has already been noted that the advantage for an arthropod parasite to involve a long-lived vertebrate is the potentially long-lasting reservoir. There are, indeed, some relatively long-lived arthropods, and in some cases their reservoir capacity is prolonged by passing the pathogen, via the egg, to the next generation. But they never approach the span of human life, and very often, especially with short-lived insects, the survival of the insect is critical, especially where a cycle of parasitic development in the insect has to occur before it becomes infective. Such development may be prolonged by cold. In temperate climates, therefore, the vector may die before becoming infective, which is one reason why diseases like malaria and yellow fever are so much more prevalent in tropical countries.

2.2 The vertebrate host: virulence; reservoirs

The host

Of the vector-borne diseases listed in Table 1, only a few (e.g. malaria, louse-borne relapsing fever) are exclusively human. Generally, there has been an extension of infections of animals (so-called zoonoses) to man; quite often this seems to have been a relatively recent evolutionary occurrence. The progress towards human disease must have been influenced by the gradual removal of man from proximity to wild animals, as he began to live in villages and towns. An example of the involvement of a new urban vector is provided by yellow fever (p. 28). Not only new vectors, but new vertebrate hosts may be involved in urbanization in the shape of pest rodents and domestic animals. Thus, plague is an endemic disease of wild rodents, which may get into domestic rodents and thence, via their fleas, to man (see p. 47). Examples of dogs as reservoirs of vector-borne diseases, which can be transmitted from them to man are Chagas' disease (p. 45) and forms of leishmaniasis (p. 35).

Changes and innovations in vector-borne disease cycles associated with incipient urbanization must have begun about two or three thousand years ago. Such changes are still taking place, though perhaps

Table 1 Pathogens and vectors of some human vector-borne diseases
* Vectors of the order Diptera, in capitals.

	Pathogens	*Diseases*	*Vectors*
'A' Viruses	WEE, EEE, VEE, etc.,	Equine encephalitides	MOSQUITOES*
'A' Viruses	CHIK, ONN	Chikungunya; O'nyong-nyong	MOSQUITOES
'B' Viruses	YF	Yellow fever	MOSQUITOES
'B' Viruses	DEN, 1, 2, 3	Dengue; haemorrhagic dengue	MOSQUITOES
'B' Viruses	RSSE, CSE	Spring-Summer fevers	Ticks
'B' Viruses	OMSK	Omsk haemorrhagic fever	Ticks
Misc. Viruses	CHF	Crimean haemorrhagic fever	Ticks
Misc. Viruses	SFN, SFS, etc.	Sandfly fevers	SANDFLIES
Rickettsiae	*Rickettsia prowazeki*	Typhus	Lice
	R. mooseri	Murine typhus	Fleas, etc.
	R. rickettsi	Spotted fever; tick typhus	Ticks
	R. orientalis	Scrub typhus	Mites
Bacteria	*Yersinia pestis*	Plague	Fleas
	Francisella tulerense	Tularaemia	Ticks, DEERFLIES
	Borellia recurrentis	Relapsing fever	Lice
	B. duttoni	Relapsing fever	Ticks
	Bartonella bacilliformis	Oroya fever	SANDFLIES
Protozoa	*Plasmodium* spp.	Malaria	MOSQUITOES
	Trypanosoma spp.	Sleeping sickness	TSETSE FLIES
	T. cruzi	Chagas' disease	Triatomid bugs
	Leishmania donovani	Kala azar	SANDFLIES
	L. tropica	Oriental sore	SANDFLIES
Filarial Nematodes	*Wuchereria bancrofti*	Filariasis	MOSQUITOES
	Brugia malayi	Filariasis	MOSQUITOES
	Loa loa	Loiasis	REDFLIES
	Onchocerca volvula	Onchocerciasis	BLACKFLIES

on a smaller scale, due to the recent uncontrolled growth of many tropical cities. This has enormously increased filariasis carried by town-living mosquitoes. Other new disease problems are initiated by the opposite process, for, when man turns back to the jungle to exploit it, he again comes in contact with vectors of diseases of wild animals.

Virulence

To succeed in establishing itself in a new host, a parasite must be able to gain access and proliferate freely. This power is described as its infectivity, and this must be carefully distinguished from the amount of harm done by the infection which constitutes its virulence. The harmful effects in a vector-borne disease can involve both man (or another vertebrate) and the arthropod vector, or it can affect either one or the other, or neither may be harmed. An example of a double virulence is provided by louse-borne typhus, which is always fatal to the louse and kills a substantial proportion of unprotected people, especially if they are not young. There are examples of pathogens harmful to man and not to the vector: they include the *Plasmodium* of malaria, the spirochaete of louse-borne relapsing fever and the trypanosome of Chagas' disease. On the other hand, a related trypanosome, *T. rangeli,* is not harmful to man, but does kill the blood-sucking bug vectors. Finally, there are various zoonoses, which seem to be harmless to both animals and vector, but become virulent when brought to man (as scrub typhus, see p. 60) or to his domestic animals (as nagana, see p. 37).

Changes in virulence during the process of adaption to man may often have gone through the following stages. A well-established vector-borne disease of animals may, from time to time, become accidentally implanted in man, due to the chance feeding of an infective vector on him. At this stage, the pathogen may not be able to overcome the natural human immunity reactions, so that it never develops a heavy infection; consequently, its virulence would be low. Suppose, however, that a change in human habits or the environment allows the vector frequent opportunities to feed on man, so that these chance infections become rather frequent. A mutation is likely to arise which will allow the parasite to establish itself in the new host. Human defences are no longer able to suppress the parasite rapidly and the result is severe virulence at this stage. Finally, however, long-established diseases (whether vector-borne or not) presumably evolve to a stage of lower virulence, with improved chances of survival of both man and pathogen.

One could find examples to support this presumed sequence of events, though it might be necessary to overlook inconvenient exceptions, and there are incidental complications such as the mere size of an infective dose, which may affect the successful infection and subsequent virulence of a disease. Also, the nutritional state of the host may alter effects of disease. Thus, starvation and fatigue may increase human susceptibility to typhus (and, indeed, cause a recurrence of a dormant infection). On the other hand, starvation of the host seems to have an adverse effect on the malarial *Plasmodium* so that epidemics are not unduly severe in times of famine. In addition to variations in the condition of the vertebrate host, there may be sudden changes in virulence of the pathogen, which cannot be readily explained by invasion of a new host.

Reservoirs

It has already been noted that for a parasite with alternate arthropod and vertebrate hosts, one advantage of the latter is its longer life with a potentially longer period of maintaining the infection; in other words, a reservoir. When the disease is of a virulent nature, this may not apply, for the animal or man is likely to die or else recover by the process of eliminating the parasite (e.g. in yellow fever). Alternatively, there is the possibility of subduing the pathogen to a dormant condition, liable to recrudesce when the host is stressed; in this way, relapses of malaria or typhus can occur after a period of years. Perhaps the most satisfactory reservoir is an animal in which the infection has reached the benign state; this may be only one of the possible hosts. There are, it would seem, wild animals (small rodents and birds) which act as reservoirs for the rickettsiae of scrub typhus and tick-borne virus infections. Again, the African game animals constitute the reservoir for the trypanosomes which cause fatal nagana when transmitted to exotic cattle or horses. Finally, one may presume that there are vector-borne diseases moderately virulent to vertebrates, which nevertheless smoulder on as zoonoses, perhaps moving about as local populations become decimated or immune. Examples could be wild rodent plague or jungle yellow fever.

3 The Vectors

3.1 Arthropod types

Primitive arthropods probably had many segments, like centipedes, but the more efficient of them fused groups of segments together for improved locomotion. The insects have grouped their segments into the head, thorax and abdomen regions; the more or less rigid thorax bearing the wings and three pairs of legs. The next most successful land arthropods, the arachnida, have a head-thorax with four pairs of legs, and an abdomen. In the mites and ticks, which will concern us in this book, the whole body tends to be fused and segmentation lost. Despite these differences in arrangements of body segments, all arthropods have some common features (see Fig. 3–1). Thus, the micro-structure of the

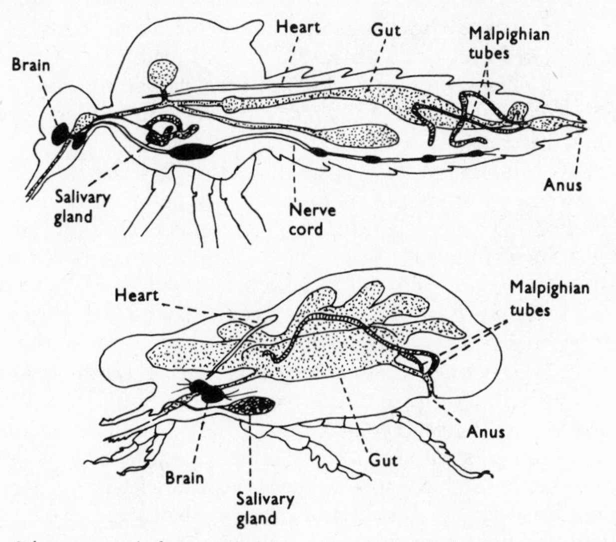

Fig. 3–1 Diagrammatic longitudinal section through the body of an insect (a mosquito) and an acarine (a soft tick) to show the disposition of internal organs

cuticle, which confers on it many valuable properties, is generally similar. There is no network of blood vessels, instead, a dorsal heart pumps blood slowly through large sinuses. This blood is not generally concerned with respiration, which in many terrestrial forms is achieved by ramifying air tubes called tracheae. The nervous system consists of a ventral pair of nerve cords with ganglia. Excretion in insects and most arachnida is by 'malpighian tubules', discharging into the rectum.

The adoption of an external skeleton (in the form of a stiff cuticle, requiring regular moults) limited the magnitude which can be attained by land arthropods. But in numbers and variety they far outstrip the higher animals. Thus, about 4500 species of mammals, 8600 birds and 5000 reptiles are known. These compare with an estimated 30 000 spiders and 50 000 mites, but even these figures are small to the number of insects, of which some 750 000 have been described.

3.2 The insects

The reason for the extraordinary prevalence and diversity of insects as compared with other land arthropods, is almost certainly their development of wings, which may have occurred well over 250 million years ago. The only drawback of wings is the difficulty of moulting, so that very early it was discovered that they were only needed in the final adult stage, for mating and dispersion. The more primitive winged insects develop them as rudiments towards the end of their juvenile (or 'nymphal') life and expand them after their last moult. Examples are crickets and cockroaches, bugs and lice. More advanced types of insect developed later with quite distinct larvae (or grubs, or caterpillars) and adults; to manage the transformation from larvae to adult a pupal or chrysalis stage was interposed. This is found in butterflies, bees, beetles, fleas, flies and mosquitoes.

In view of the staggeringly large number of insect species, it is hardly surprising that most people are familiar with many of the main types, so that the natural orders (lepidoptera, coleoptera, etc.) can be described by well-known common names (moths, beetles, etc.). Nevertheless, it is obvious that we are only aware of a tiny fraction of these unnoticed hordes of minute creatures. Apart from a few showy insects, such as butterflies, and house-haunting species like the housefly, the insects we notice are mainly ones which afflict us as pests. A recent estimate of the more important pests gives a figure of 4500. Although this is only a tiny proportion (well below 1%) of known insects, the harm done by these pests is quite substantial. The majority are important in agriculture, attacking field crops of foods (cereals, vegetable, fruit), fibres (cotton), beverages (tea, coffee, cocoa) and other crops (sugar, tobacco, spices). Forest pests blight growing trees. Veterinary pests attack cattle, sheep and other domestic animals. Food products in stores and stacked timber are damaged. In the public health field, those directly harmful by stings and bites are much less serious than species which carry disease, which are the subject of this book.

Although insects require the same basic foods as other animals (carbohydrates, protein, fat) they obtain them from an extraordinary variety of sources: green leaves, seeds, wood, decaying vegetation, honey, blood. To feed on such substances requires very different feeding apparatus.

It seems likely that the mouthparts of the primitive insects were adapted for chewing, like those of the cockroach shown in Fig. 3–2. Analagous parts can be recognized in various other orders of which the members chew solid food. These include a vast number of both

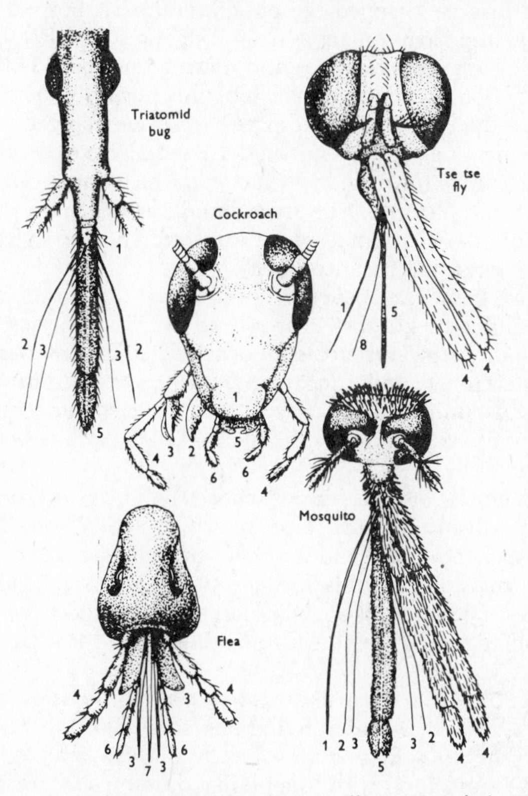

Fig. 3–2 Heads of some blood-sucking insects illustrating their mouthparts (some displaced, to show them). **Key**: 1, Labrum; 2, mandibles; 3, maxillae; 4, maxilliary palps; 5, labium; 6, labial palps

primitive and advanced insect types; dragonflies, cockroaches, stoneflies, termites, stick insects, grasshoppers, beetles, ants, wasps and bees. The primary elements are the following: a *labrum* (or upper lip); a pair of tooth-like *mandibles*; a pair of *maxillae*, with a pair of terminal processes for triturating food; a labium (or lower lip) showing traces of a paired structure. Attached to the maxillae and also to the labium, are pairs of jointed, finger-like palps. The most essential element is the pair of toothed mandibles, which is also found in larvae of the beetles and those of butterflies and moths (though the adults of the latter have abandoned the chewing habit and their mouthparts are highly modified for sucking nectar from plants). It is among these chewing insects that are found all the main agricultural pest insects. They have, however, little or no importance in public health and will not concern us further.

The astonishing moulding power of evolution has, in various other groups of insects, changed the basic mouthparts beyond recognition. The most important change from our present viewpoint is their modification for piercing tissue and sucking out fluids, either the sap of plants or blood of other animals. For this purpose, the mandible and maxillae of the primitive mouthpart have been changed into long thin stylets for piercing the host's tissues. These stylets are curved in section, and fit together to form a narrow tube, up which liquid can be sucked by a pump in the pharynx. The stylet bundle is held in a groove on the jointed labium which guides it into the wound. The stylet bundle and labium are known as the 'proboscis'.

There are good reasons for supposing that this has occurred several times, in different groups. Probably the earliest of these adaptations resulted in the large and successful order called the *hemiptera*, or bugs. It has also developed, quite separately, in the sucking lice and the fleas. The latter, unlike the other two orders, undergo complete metamorphosis and it is only the blood-sucking adults which show this modification.

The remaining insects, with which we shall be concerned, belong to the highly advanced order diptera, which shows some evolutionary trends within the order and a variety of well-known types. The most primitive division of this order (the *nematocera*, which includes mosquitoes, biting midges, blackflies and sandflies) have adopted a rather similar arrangement, with mandibles and maxillae converted to stylets.

The highest, or at least the most specialized, division of flies is the *cyclorrhapha*, which includes numerous small forms like the fruit fly *Drosophila*, the housefly and blowflies, has largely adopted the habit of licking up liquid food with tongue-like mouthparts. For this purpose, the labium has been modified into an organ with fleshy pads covered with fine canals leading to the mouth orifice. All traces of mandibles and maxillae are lost, except the maxilliary palps, which are reduced to a single joint. One could hardly suppose that this apparatus could be re-converted into a piercing and blood-sucking organ, but this has happened in tsetse flies, stable flies, etc., and also in the peculiar parasitic forms allied to the sheep ked and forest fly. For this purpose, the elastic labium with its spongy lobes has been converted to a horny spike, which pierces the skin just as well as the stylets of the nematocera. This evolution probably happened among flies which hover around animals and tend to lick up blood from small scratches. In this advanced order of diptera both sexes take meals of blood.

If the beetles, moths and bugs provide most examples of agricultural pest insects, there is no doubt of the pre-eminence of the diptera in medical entomology. Outside this order, blood-sucking forms are relatively few and tend to be 'degenerate' parasites. That is to say, they have become so closely associated with their hosts, that they have mostly dispensed with the need for wings. In lice and fleas, no trace of wings is left. Bed bugs have tiny vestigial scales. The large tropical blood-sucking bugs of the family triatomines have functional wings it is true, but they are not vigorous or persistent flyers. In contrast, the mosquitoes, midges and sandflies are moderate flyers, with a range of a mile or two. Blackflies can travel a score or more miles by air. Horseflies, tsetse flies and stable flies are rapid and powerful in flight. Only in the highly specialized group called pupipara has the sluggish, host-dependent habit introduced a wing reduction. Apart from the many blood-sucking dipteran forms, therefore, their mobility gives an extra reason for their efficiency in spreading certain diseases. The frequency with which they are involved can be noted from the data in Table 1.

3.3 The acarina, or mites and ticks

The size range exploited by the acarina is considerably below that of the insects. The vast majority are between 0.2 and 2 mm long in the adult stage; and for this reason we are little aware of their immense numbers and often strange diversity. The very great majority are free-living inhabitants of soil or surface vegetation, feeding on plant or animal remains, or on faeces. Some feed on living plants forming galls in the process; others enter into various relations with insects or larger animals, sometimes merely employing them as modes of transport, but often becoming parasitic in various ways. There are many skin parasites of vertebrates, some only feeding on secretions, others causing injury by burrowing into the skin (as the mange mites and the human scabies mite). There are a few mites which have adopted the habit of feeding on the blood of birds and mammals (e.g. the red poultry mite). The larval stage of the harvest mite similarly feeds parasitically in higher animals, though it seems to suck out serum rather than blood. We shall meet its tropical relatives later since they are vectors of scrub typhus. The ticks are, of course, the giants of the order, some of them reaching a length of 2 or 3 cm. All of them feed on the blood of mammals, birds, or reptiles, and quite a number are important vectors of diseases of man or domestic animals.

The classification of acarina is much less straightforward than that of insects, and a number of different systems have been suggested. One of the more modern arrangements divides them first according to a basic constituent of the cuticle and thereafter according to the position of the *spiracles* (or openings of the respiratory system). These are often rather difficult to detect, but quite a number of groups have a 'typical' general appearance for those who study them. Since wings were never developed there was no occasion for the introduction of complete metamorphosis. Most acarina lay eggs, from which emerge larvae with only three pairs of legs. After subsequent moults there are one or more four-legged nymphal stages before the adult appears. The larval habits may differ from those of other stages; but nymphs and adults usually have similar feeding arrangements.

The principal mouthparts are paired *palps* and *chelicerae* (which may be pincer-like), with a ventral, central *hypostome*. In the ticks, with which we are mainly concerned, the mouthparts are adapted for taking prolonged blood meals. The chelicerae are provided with lacerating processes and the enlarged hypostome bears recurred teeth, to anchor the tick in the wound. These various units normally project forward together; but, in feeding, the stumpy palps are splayed sideways, while chelicerae and hypostome penetrate the flesh. In the so-called 'hard' ticks, the mouthparts project forward in front of the body. In the other group, or 'soft' ticks, the mouthparts are hidden by an anterior fold of the body. The two groups are rather distinct in habits as well as anatomy, and these differences will be briefly discussed when we consider them as disease vectors (p. 55).

4 Mosquito-borne Diseases

Though mosquitoes belong to the most primitive branch of the advanced order diptera, in one respect they are highly specialized. Other blood-sucking members of the order have relatively short mouthparts, which make small pit-like wounds into which blood seeps and is imbibed; this is described as 'pool-feeding'. Mosquitoes, on the other hand, have long slender stylets, which probe the flesh until they find a blood capillary, from which they draw blood very rapidly.

Disease vectors are scattered in numerous species in different branches of the mosquito family, the culicidae. The main divisions are the relatively homogeneous anophelines and the more diverse culicines. These two can be distinguished in each stage of the life cycle. Anopheline eggs are cigar-shaped with side floats, whereas culicine eggs have no floats. Anopheline eggs are scattered over the water. Culicine eggs may be laid in raft-like masses (*Culex*), or laid on damp surfaces to await subsequent flooding (*Aedes*), or laid in cushion-like masses under leaves of water plants (*Mansonia*). Larvae of anophelines lie immediately under the water surface and parallel to it; those of *Culex* and *Aedes* hang down from a respiratory siphon, the tip of which is anchored in the surface film; those of *Mansonia* live submerged and draw air from siphons thrust into air cavities in the stems of aquatic vegetation. Adults of anophelines rest with their head, thorax and abdomen in a straight line, inclined at an angle to the surface, whereas culicines adopt a more hump-backed position (Fig. 4–1).

A bare outline of the life cycle cannot convey the great variety of ecological variation. Various species prefer particular types of breeding site. These include tiny seepages, rock pools, water in plant axils or rot holes in trees, ponds, lake margins, streams and marshes. The water may be fresh, brackish, or polluted; in clear sunlight, or in shade. One cannot easily generalize here as to anopheline or culicine preferences, except perhaps to say that the former mostly require unpolluted water, while some culicines occur in polluted sources. The rate of development of mosquitoes, like that of all arthropods, is dependent on climate. In hot countries, the life cycle is completed in a week or so, and breeding is continuous, though often augmented in rainy seasons and reduced in dry ones. In temperate climates, however, life cycles usually require several weeks even in summer, and in winter the insects hibernate, either as eggs, larvae or adults. All kinds of warm-blooded and some cold-blooded land animals may serve as sources of blood for the females. Generally, mosquitoes have preference for feeding on certain hosts, but they are not rigidly specific. Obviously, those which are important in the transmission of human infections are the ones liable to feed readily on man.

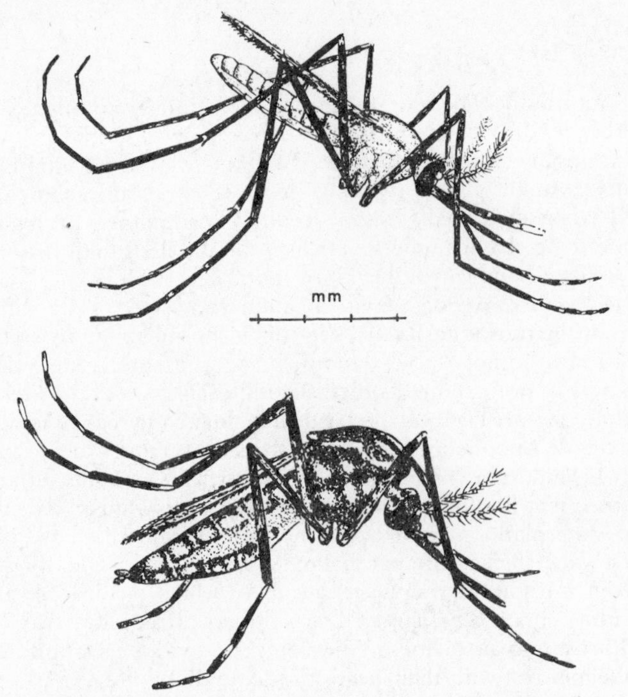

Fig. 4–1 Mosquitoes. **Above,** *Anopheles gambiae*, a major African malaria vector. **Below,** *Aedes aegypti*, vector of urban yellow fever

4.1 Malaria

History and importance

Biological evidence suggests that man's ancestors have been subject to a form of malaria since pre-human times, so that it would not be surprising to find mention of it in very early historical records. Many early references (Old Testament and ancient oriental works) are rather vague and uncertain, but since the clear descriptions of Hippocrates (460–377 B.C.) it has been possible to trace the rise and fall of malaria, in Europe at least. Resurgence of malaria was often associated with a collapse of civilization but cause and effect are not easy to establish. Does civilized life decay under the influence of the disease? Or does the disease increase when barbarian invaders disrupt settled agriculture and drainage systems are neglected or destroyed?

At the present time malaria is almost certainly the most important arthropod-borne disease, being responsible for an immense amount of human illness and early death. Though more common and generally much more intense in tropical countries, malaria can be severe in the sub-tropics and extends into the temperate zone. In Chapter 1 it was noted that the vast scale of world malaria and the improvement due to human effort (see also Fig. 4–2). Twenty-five years ago the annual incidence was estimated as about 300 million, with about 3 million deaths; today these figures are in the region of 50 million with under a million deaths. This encouraging progress has been largely due to the use of that currently unpopular chemical DDT, used as a residual insecticide against the mosquito vectors. Unfortunately, however, it cannot be anticipated that this reduction of malaria will continue, because the areas which remain are very heavily affected and present a much more severe challenge than the peripheral zones which have been reclaimed from the disease.

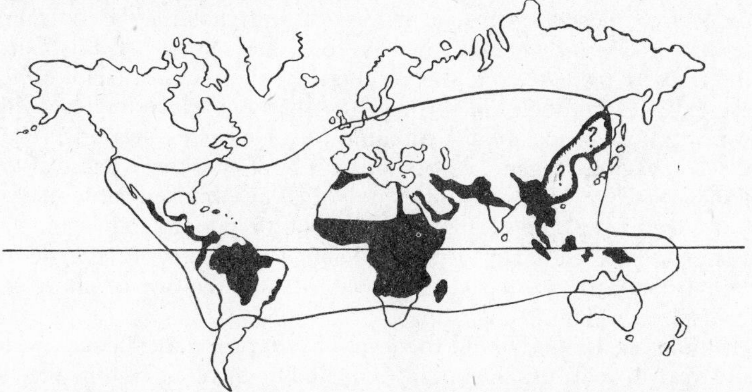

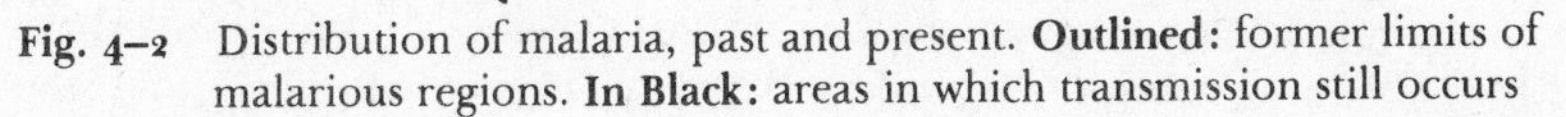

Fig. 4–2 Distribution of malaria, past and present. **Outlined:** former limits of malarious regions. **In Black:** areas in which transmission still occurs

Nature of the disease

Human malaria is a term covering four rather closely related diseases due to different species of parasitic protozoa in the genus *Plasmodium*. These parasites can in some instances, be transmitted to monkeys, and other species responsible for analogous infections in chimpanzees may be transferred to man. It seems, however, that this scarcely ever happens in nature, so that malaria is a specifically human disease with man as the sole reservoir of infection. All the forms of malaria involve characteristic periods of chills and fever separated by periods of remission; they differ, however, in periodicity, malignancy and liability to relapse. 'Quartan malaria' has a four-day cycle, comprising a day of fever, two days of remission and another bout of fever. The other forms are described as 'tertian', with three-day cycles; that is, a single day of remission between two days of fever. Of the two most common forms of tertian malaria, one is described as 'benign' and the other as 'malignant' because the latter involves more danger of fatality. In some cases, people may be infected with two or three of the parasites, most commonly those of benign and malignant tertian malaria.

As regards distribution, malignant tertian requires a warmer temperature for transmission and is thus confined to the tropics and sub-tropics. Benign tertian, though common in the tropics, is able to spread into temperate regions as far north as England and Siberia, where the parasite has developed a long latent period in man, in order to pass through the winter when no mosquitoes are available for further transmission. Quartan malaria has a wide geographic range but is distinctly patchy in distribution.

A degree of acquired immunity is conferred by attacks of malaria; it is not of the same protective value as the immunity from some other diseases, and is sometimes described as premunition. In areas where people are exposed to constant reinfection with malaria, the indigenous population acquires more or less tolerance. Some of this partial immunity is passed from the mothers to their unborn children by passage of immunoglobin through the placenta. Nevertheless, a high proportion of young infants succumb to their early infections, so that the tolerance of the general population is bought at the expense of high infant mortality. This situation is described as hyper-endemic malaria. It can readily be appreciated that susceptible people arriving in such hyper-endemic countries are very liable to succumb to the disease, unless adequately protected. Such indeed, was the fate of many early European travellers in the tropics.

In other malarious zones, the level of infection may fluctuate widely from year to year, usually because of climatic or other factors affecting the prevalence or longevity of the mosquito vectors. In such cases, there may be long periods with low rates of transmission, so that the inhabitants tend to lose their premunition. When a season favouring transmission arrives, the result is a rapid spread of severe malaria among susceptible people; in other words, an epidemic.

The parasites

No fossil record illuminates the evolution of parasitic protozoa, but much can be inferred from the biology of more primitive types. It is currently believed that the remote ancestors of *Plasmodium* were gut parasites of a vertebrate, probably an aquatic one.

The parasite cycle is as follows. Infective forms ('sporozoites') are injected into the blood stream with the saliva of the mosquito. These travel to the liver, invade the cells and pass through one or more asexual cycles, involving division into daughter cells, which are released on rupture of the host cell. So far, the infected person feels no symptoms, and at this stage is described as the incubation, or pre-patent period. In warm countries, this usually lasts for about a week, but sometimes in temperate countries, the pre-patent period may last from autumn until the following summer. Eventually, some of the parasites invade the blood stream and enter red corpuscles (erythrocytes) in which they undergo repeated asexual reproduction. These erythrocytic cycles tend to be synchronized at 48 or 72 hours, according to the species of *Plasmodium*. At each cycle, vast numbers of blood cells are ruptured, releasing toxins which cause a paroxysm of fever. Thus, the tertian or quartan forms of malaria are related to the length of the synchronized reproductive cycles. Eventually, the precursors of sexual gametes are formed, which also invade blood cells, but do not fully develop until swallowed by a mosquito. In the insect's stomach, the male gametes fuse with female forms and the fertilized 'ookinete' then burrows through the stomach wall and forms a kind of cyst on the outside. In each of these, numerous, spindle-shaped sporozoites develop, burst out, and make their way to the mosquito's salivary glands, ready to invade a new host.

Meanwhile, patients who do not succumb, produce antibodies which suppress the erythrocytic cycles of the parasite. This immunity reaction is mediated by the spleen, which becomes enlarged. The parasites of tertian malaria are more dangerous than the other forms, but if they are suppressed in this way, a cure is normally effected. The other parasites, however, may persist, either in the liver or at extremely low levels in the blood. Some time later, when the antibodies in the blood may decline for some reason, a new attack or malaria relapse will occur. This can happen a long time after an infection of quartan malaria, five or eight years is not an uncommon period and over forty years has been recorded. With benign tertian malaria, however, the relapses rarely occur after more than a year or two.

Phases of the *Plasmodium* life cycle in man are thus of interest in relation to pathogenicity. In the mosquito, the parasites do not seem to have any serious adverse effect. But this part of the life history is of prime importance in relation to the prevalence and geographical distribution of the disease. The essential point is that during its cycle in the mosquito, the parasite is exposed to ambient temperature, on which its rate of development is closely dependent. In tropical temperatures, this requires only a week or two, but the time is progressively lengthened in cooler climates. Since mosquitoes are not long lived, they may not survive long enough to transmit the parasite. In the tropics, on the other hand, not only are mosquitoes plentiful but parasites mature in them much more quickly.

The anopheline vectors

Some 350 species of *Anopheles* are known, some widely distributed, others localized. As might be expected, considerable numbers of species are found in some parts of the world and few in others; Britain has five species, the Malaya Peninsula about sixty. By no means all of these are important vectors of malaria, usually only one or two species in any given locality are primarily responsible for transmission, and the world total of these major vectors is not much more than a score.

We should now consider the characters which make an efficient vector. To some extent, different anopheline species vary in their susceptibility to infection, but more important reasons relate to the feeding habits and ecology of the mosquitoes. Female mosquitoes are by no means specific in their food requirements but they do show some distinct preferences. In order to acquire an infection and later transmit malaria, a mosquito must feed at least twice on man, so that readiness to feed on man is a fundamental characteristic of a malaria vector.

The majority of serious vectors feed at dusk or during the night, commonly on sleeping people. This usually involves entering a human dwelling to do so. As might be expected mosquitoes tend to seek blood meals not too distant from their breeding sites, usually within a kilometre or so. Consequently, the proximity of breeding places suitable for a vector species involves a risk of malaria. Sometimes human operations create such sites. For example, removal of clay for building leaves borrow pits; growing of rice, or clearing of jungle may offer improved breeding places and lead to 'man-made malaria'. After the blood meal, female mosquitoes which have fed on sleeping people often rest inside the dwelling to digest their meal. A considerable proportion will leave at dawn each day and if the blood meal is digested, the mosquito will seek a breeding site to lay her eggs. She is then ready for another meal, and unless the mosquito is a hibernating species facing a temperate winter, the process will be repeated continuously until death. The length of the cycle depends on temperature, and in tropical climates is often as short as two days.

It has been noted that the malaria parasite needs a week or two to reach the infective stage in the mosquito. In order to be an effective vector, the insect must obviously survive beyond this point. This is easily accomplished by mosquitoes kept under equable conditions in captivity, but in the field, with the constant need for feeding and oviposition, hazards are much higher. In recent years ways have been found of estimating the adult ages of wild mosquitoes. Results suggest that the death rate is fairly constant (often around 10% per day) and depends on environmental hazards and not on ageing.

Measures against malaria

DRUGS AGAINST THE PARASITE Until the present century almost nothing could be done to combat malaria except to dose the sufferers with quinine, which was first isolated from cinchona back in 1820. Though the quinine alkaloids destroy malaria parasites in the blood, their prolonged administration (which may be demanded in highly malarial places) may have harmful side effects leading, for example, to the dangerous complication known as blackwater fever. In the 1930's, the first effective anti-malarial drugs were introduced and since then, many more have been prepared, with different types of action. It is evident that the physician has a battery of medicines for cure or prophylaxis. The need for careful choice and judicious economy in these drugs is emphasized by the emergence in several areas of strains of *Plasmodium* resistant to some of them.

The existence of relatively cheap and safe anti-malarial prophylactics has made the tropics much safer for highly susceptible visitors. Also it would now seem theoretically possible to eradicate malaria completely from large areas by mass medication. Unfortunately, it is virtually impossible to rely on regular self-dosing by all inhabitants of under-developed countries. Attempts have been made to ensure universal treatment, by adding anti-malarial drugs to salt (in the way fluoride is added to drinking water in some countries for dental prophylaxis). This has met with limited success and could well be incorporated into integrated campaigns of the future.

MEASURES AGAINST THE VECTOR From the discovery of the part played by the mosquito in transmission of malaria, there have been continual attempts to reduce or eliminate the disease by attacks on the vector. In the first three decades of the twentieth century, two main methods were used; the larvae were destroyed by oiling the breeding sites and attempts were made to eliminate such sites by drainage. In the 1930's schemes of breeding-site reduction were elaborated into the technique of 'species sanitation'. The biology of each vector species was studied with care to find out its breeding requirements. Then the locality was modified to alter or eliminate favourable sites, not merely by drainage. Shade-loving species were deterred by cutting down vegetation along streams, sun-loving forms were deterred by planting shade trees. Vector species which bred in salt marshes were eliminated by dykes to prevent infiltration of sea water. All such measures were admirable applications of scientific knowledge and, furthermore, they gave long-lasting reduction of the vector species. But they had one drawback, they were expensive of labour and resources, so that only rather small areas could be protected in this way.

The introduction of DDT and other residual insecticides opened a new era by producing a control measure that took advantage of a weak link in the life cycle of nearly every vector. As has been noted, most of these enter human dwellings to feed and thus contact the walls on which they usually rest. The application of long-lasting insecticide to such walls took a continual toll of the actual vectors and so, by expenditure of a relatively small sum, inhabitants of tropical dwellings could be substantially protected from malaria for a season. The treatment of houses and huts was simple, so that it could be done by mobile teams operating in remote rural districts. The results in the first decades after the Second World War were almost universally encouraging. It was this method of 'residual house spraying' which suggested the possibility of global eradication of malaria, sponsored by the World Health Organization in the 1950's and 1960's. To eradicate malaria, it is not necessary to exterminate the vectors (a virtually impossible task), but merely to reduce numbers until transmission ceases and maintain this situation until all human infections die out. Campaigns undertaken by various countries were co-ordinated by the W.H.O., which provided much technical assistance. The basic plan, in each country, comprised a series of phases: preparatory, attack, consolidation and maintenance. The admirable results of this complex undertaking have already been mentioned briefly and these are further illustrated in Fig. 4–2. The difficulties in the way of further progress are basically twofold: (i) lack of adequate funds and trained personnel, especially in view of other demands on both, in many tropical countries. (ii) Certain anopheline vectors which are unduly refractory, either because they have developed strains resistant to the insecticides (see p. 65), or because their habits enable them to avoid the insecticide.

4.2 Filariasis

History and importance

Filarial infections transmitted by mosquitoes have doubtless afflicted people in the tropics since pre-historic times. They are not acute or lethal diseases and their main importance is the secondary complication of elephantiasis. Visitors to the tropics will notice a proportion of native inhabitants with enormously swollen legs or other parts of the body (see Fig. 4–3). These sufferers represent only a fraction of the vast numbers actually infected with the parasites, which are estimated at over 250 million, spread round the tropical zone. Moreover, the incidence is still growing, largely due to the rapid growth in recent decades, of unsanitary slums round tropical cities, where open drains provide breeding sites for the vector. The elimination of this unpleasant disease poses problems at least as difficult as the eradication of malaria; and since it lacks the lethal urgency of some other diseases, it has until recently provoked much less vigorous counter measures.

Nature of the disease

Mosquito-borne filariases are morbid conditions due to nematode worms of the genera *Wuchereria* or *Brugia*. *W. bancrofti* is particularly important; it occurs throughout the tropics and seems to be restricted to man. *Brugia* spp. occur in South-East Asia and include forms largely restricted to animals. The life cycles of both are similar in general. The adults are thread-like worms a few centimetres long, which live in lymphatics and connective tissue. The fertilized females produce numerous immature larvae about 0.25 mm long, known as micro-filariae. These alternate between the peripheral blood stream and the

Fig. 4–3 An East African man suffering from elephantiasis of the left leg, due to filariasis. (Photo courtesy of Professor G. S. Nelson)

lungs or the main arteries. They cannot develop further until swallowed by a vector mosquito and their prevalence in the peripheral blood usually shows periodicity which corresponds to the biting habits of particular vectors. Those of *W. bancrofti* feed mainly at night, so that its micro-filariae are most abundant between 10 p.m. and 2 a.m. In the Pacific, however, a sub-species exists, of which the vectors are daytime feeders, and the periodicity of the micro-filariae is correspondingly reversed.

The initial effects of filarial infection in man are slight. During the long period before the worms become adult, transient fevers with local pains and headache may occur. These are exacerbated when the worms become mature, and are apparently due to toxins released by them, since the micro-filariae seem to be harmless. In about 10 to 15% of cases where the worms are numerous, they eventually cause obstruction at the lymph nodes; as a result, the vessels gradually swell and distend the flesh to enormous extents, causing elephantiasis. Commonly the legs are involved, together with the scrotum in men and sometimes the breasts in women. Scrotal swellings of 10 to 20 lb. are common and up to 50 lb. by no means rare. The inconvenience and discomfort can be imagined.

The parasites

Many nematodes live saprophytically in the soil and some, known as 'eelworms', parasitize plants. Single-host animal parasitism is represented by the round intestinal 'worms' of domestic animals, with oral infection and dispersion through faeces. The 'filaroid' type of worm, invades tissues of the vertebrate and is transmitted by blood-sucking diptera, the mosquito being an advanced form. It should be noted that throughout this entire evolution, sexual reproduction occurs only in the vertebrate, which is therefore the definitive host.

In the mosquitoes' stomach, some of the micro-filariae are digested and a few defaecated. The remainder manage to get into the body cavity and eventually reach the thoracic muscles, which they penetrate. Here they grow and pass through one or two moults before becoming ready to reinfect a vertebrate. The time needed for this maturation ranges from two to six weeks according to temperature. Mature worms move about in the body and accumulate in the proboscis. When the mosquito feeds, the worms burst out of the tip of the proboscis and enter the vertebrate, usually at the bite puncture. They soon proceed to the lymphatics where further larval development takes place, producing adults after three months or more. These may live for one or two years.

The mosquito vectors

Various filarial worms can be transmitted by a considerable variety of arthropods, though the range for each species is rather more limited. The mosquito-transmitted human filaria can involve many species of anophelines and culicines, but some of these are particularly important. The factors deciding vector capacity (as with malaria) are both physiological and ecological. The chances of worms establishing themselves in a mosquito after a blood meal depends on their surviving antagonistic physiological reactions of the mosquito (digestion in the stomach; encapsulation in the muscles). This not only varies between species of mosquito, but even between strains of a single species. In addition to the effect of the mosquito on the worms, we must consider the converse. Heavy worm loads impede digestion and may harm the mosquitoes' stomach; also the relatively large worms among the flying muscles probably impede flight.

The ecological factor mainly relates to frequent man-biting, and where a very suitable mosquito vector occurs, the periodicity of the micro-filariae in the blood stream is likely to conform to its habits. Thus, the pre-eminent vector of *W. bancrofti* through much of its range and especially in town and cities is the night-feeding *Culex pipiens fatigans*; but day-feeding *Aedes* spp. transmit the Pacific race. Various mosquitoes transmit *Brugia* including *Mansonia*, the culicine with larvae which obtain oxygen from aquatic plants.

Treatment and control of filariasis

Both measures against the parasites and the vectors are beset with difficulties. Drugs can be used against the worms, notably Hetrazan, which destroys adult worms and eliminates micro-filariae. But developing larvae are unharmed, so treatment must be prolonged or there may be relapses. Furthermore, the drug is liable to uncomfortable side-effects so that symptomless infected people resent being dosed. It is therefore difficult to conduct sustained mass-medication campaigns to eradicate the disease.

Control of each of the different types of vector presents special problems. Perhaps the most important is the urban *Culex p. fatigans* which breeds in polluted drains. Sanitation is poor in rapidly growing cities throughout the tropics and this mosquito abounds in them. Attempts to control it by insecticides are frustrated by the readiness with which this species develops resistant strains. Currently investigations are in progress on ingenious modern alternative methods such as the sterilization of local mosquitoes by the release of genetically incompatible males, but progress is slow and uncertain. The ultimate objective should at least include better drainage and houses with covered sewage disposal systems.

4·3 Arboviruses

Viruses transmitted from one vertebrate to another by insects or acarines are known as arboviruses (short for arthropod-borne viruses). In the past decade the numbers of examples known has risen from about 70 to over 200. Three-quarters of them are transmitted to mosquitoes and a few by other biting diptera, the remainder being spread by ticks (see p. 62). In the vertebrate, the virus may invade many different tissues, with corresponding unpleasant symptoms. The most serious forms are the haemorrhagic fevers and the encephalitides. In the former, the virus attacks capillary walls, increasing their permeability and causing bleeding from the gums, nose, uterus, lungs and kidneys.

Encephalitis virus, on the other hand, invades the central nervous system, with effects rather similar to poliomyelitis. Less malignant arbovirus infections cause severe pain in back and joints and are described as break-bone fevers, the best known being dengue. All arboviruses must, of course, invade the blood to some degree for transmission by blood-sucking arthropod vectors. The levels of viraemia reached, however, varies in different vertebrate hosts, and so does the susceptibility to infection of different mosquito vectors. These two factors, as well as the prevalence and feeding habits of the vector, largely determine the infection cycles of different arboviruses.

History and importance

Most mosquito-borne arboviruses are tropical, and Europeans knew little about them until the extensive exploration and colonization during the nineteenth century. Then they encountered the appalling epidemics of yellow fever in tropical towns of Africa and the less dangerous, but unpleasant, attacks of dengue. The major urban epidemics of yellow fever occurred at the end of the nineteenth and the early decades of the twentieth century.

Yellow fever occurs in South and Central America, from southern Brazil to British Honduras, and in Africa in a belt about 10° S. and 15° N. of the equator. It has never reached Asia and since suitable vectors exist there, health authorities vigilantly monitor transport (especially aircraft) which might introduce infected mosquitoes. There are two types of disease incidence related to different transmission cycles, the urban and jungle forms. The urban disease, which is potentially epidemic, involves only man and urban mosquitoes. Jungle yellow fever, which tends to be sporadic rather than epidemic, depends on fortuitous infection of virus from a permanent reservoir in wild monkeys, among which the infection is transmitted by forest mosquitoes. The disease is often fatal to South American monkeys, though not to African ones, but both acquire permanent immunity after recovery, as man does. Since the virus does not persist in such cases, the reservoir consists in infections wandering about over large areas, constantly involving susceptible animals. Human involvement comes about mainly by people, especially tree cutters, intruding into forests and being bitten by the mosquitoes which normally feed on monkeys. If such a person travels to a town before falling ill, he may infect the urban vector if it is prevalent, and thus initiate an epidemic.

Once the mosquito transmission was established (about 1905) control of the urban vector was possible, and later, immunization procedures became available to protect susceptible individuals. Progress in recent years, both by vector control and mass inoculation, has virtually eliminated urban epidemics in the Americas and greatly reduced their severity in Africa. World incidence of the disease as reported to W.H.O.

in the past twenty years, fluctuates from about 100 to a few thousand. Many more undetected cases must occur, however, and as it is unlikely that the disease will ever be eradicated, constant vigilance is necessary.

Dengue and other similar complaints are still common in many tropical countries. In 1972, half a million cases occurred in Colombia alone. In the last decade, a virulent form, known as haemorrhagic dengue, has developed in South-East Asia. The numbers involved, however, are not enormous. Among other mosquito-borne arboviruses are encephalitides, which have caused unpleasant outbreaks in America and the Far East in the past forty years.

Nature of the diseases

YELLOW FEVER This is the most important of the haemorrhagic group, though perhaps the signs and symptoms are not entirely typical. It affects man and other vertebrates, especially monkeys. In man the effects range from very mild to sometimes fatal, young people generally being the most tolerant. After a short (3–6 day) incubation, severe cases suffer from sudden onset of fever, headache, pains in loins and legs, black vomit and diarrhoea. The liver is severely affected and a jaundice is caused in many patients, especially in fatal cases, hence the name of the disease.

HAEMORRHAGIC DENGUE It occurs in South-East Asia and seems to be prevalent recently, apparently due to double infections of virus. Severe urban epidemics with thousands of cases and some hundred deaths have occurred. Virtually all cases are among native children, white people being unaffected for some reason. It produces typical haemorrhagic effects and is rather malignant with a mortality rate of about 7%. Largely an urban disease, it is transmitted by *Aedes aegypti*.

ENCEPHALITIDES The American diseases in this group include, Western equine encephalitis, Eastern equine encephalitis, St. Louis virus and Venezuelan equine encephalitis (often abbreviated to WEE, EEE, etc.). It has been established that these normally affect birds, being transmitted mainly by culicine mosquitoes which feed on them. Horses and people can be infected by bites of the same mosquitoes, but in each case, no further transmission occurs. The human disease is not fatal, but tends to leave spastic effects similar to those due to poliomyelitis. In the Far East, a similar infection is known as Japanese B encephalitis.

BREAK-BONE FEVERS Dengue is a non-lethal, febrile illness characterized by a rash and intense pains in the joints. In recent years other diseases with similar symptoms have been observed in Africa and Asia. Several are named after local dialect words for break-bone or break-joint, as *Chikungunya* and *O'nyong-nyong*, or *O'bili bili*. No reservoirs other than man are known, though it is possible that some unknown ones may exist.

The pathogen

The arboviruses are a heterogeneous group of RNA-containing viruses, spherical or rod-shaped, of about 120–125 μm. They have been classified serologically, according to the number of antigens in common, originally as Groups A, B and C, and later according to characteristic members. The types of disease caused do not, however, relate closely to the virus groups (see Table 1).

On inoculation by mosquito bite into a susceptible host, the virus multiplies rapidly. This 'incubâtion period' with yellow fever is only 3–4 days, or with dengue, 5–8 days. In non-fatal cases, the viraemia is suppressed by antibodies in a few days, so that only a rather short time is available for infection of a further vector. After multiplying in the cells lining the mosquito's stomach, the virus spreads to various other tissues, especially of the nervous system and the salivary glands. The period before the insect becomes infective varies considerably in different species, but in all cases it is temperature-dependent. Thus, yellow fever virus in *Aedes aegypti* requires thirty-six days at 18°C but only seven days at 37°C. The mosquito does not seem to be harmed by the virus infection and remains infective for the rest of its life.

The vectors

The great majority of mosquito-borne arboviruses are transmitted by culicines. Apart from intrinsic susceptibility and feeding preferences, the species involved depend on their ecology and behaviour. Yellow fever virus is spread among forest monkeys by species which inhabit the forest canopy, species of *Haemogogus* in South America, *Aedes africanus* in Africa. In both continents the highly domestic *Aedes aegypti*, which breeds near and frequently enters human dwellings, is a potential vector of urban epidemics. This mosquito is also the usual vector of dengue throughout the world, and of chikungunya in East Africa. O'nyong-nyong, however, is exceptional in being transmitted by anopheline mosquitoes.

The encephalitides are commonly maintained in the bird reservoir by mosquitoes of the genus *Culex*, which will sometimes feed on mammals, including man.

Disease control

YELLOW FEVER Prior to the discovery of the jungle reservoir in the 1930's, it seemed that this was a strictly human disease. Since the vector was not difficult to control and since human cases do not long remain infective, the total eradication of the disease seemed quite feasible. The recurrence of urban epidemics, introduced from jungle sources, was a severe set-back. It now seems unlikely that the virus can ever be completely eliminated from monkey reservoirs in tropical forests, which consequently must be regarded as a permanent threat. Around 1950 an alternative strategy was adopted in the Americas with a co-ordinated effort to eradicate the urban vector, *Aedes aegypti*, from all towns and cities. DDT was used systematically against both adults and larvae in and around dwellings. Some fifteen years later eradication was claimed by seventeen countries in the hemisphere, but resistance to DDT and other insecticides has developed and with other difficulties it now seems unlikely that complete eradication will be achieved. Indeed, there are some re-invasions. Alternative insecticides are still available for control among phosphorus and carbamate compounds. These need to be preserved as weapons to combat any future epidemics.

Vector control is supplemented by use of immunizing vaccine on a large scale. In those African endemic areas formerly under French control, for example, over 60 million doses were given between 1940 and 1960. In addition to these measures, active surveillance for cases of yellow fever is maintained by many tropical authorities.

OTHER DISEASES Control of dengue outbreaks depend largely on measures against *Aedes aegypti*. In South-East Asia, where the dangerous haemorrhagic variety has been prevalent, DDT-resistance is a difficulty. Measures to cope with urban epidemics include spraying of ultra-fine malathion aerosol from aircraft.

5 Diseases transmitted by Diptera other than Mosquitoes

Four examples of disease-carrying diptera are discussed in this chapter; two (blackflies, sandflies) from the primitive nematocera branch of the order and two (tsetse flies, houseflies) from the advanced cyclorrhapha branch. Among the intermediate brachycera, some less important vectors occur in the biting horsefly family. In West Africa, 'redflies' transmit an unpleasant but not dangerous filarial disease known as Loa loa. In the U.S.A., rather similar 'deer flies' take part in sporadic transmission of the bacterial disease, tularaemia.

5.1 Onchocerciasis, River Blindness and Blackflies

Importance and distribution

Onchocerciasis is a disease due to a filarial nematode and is a gradual, chronic infection, rather than an acute dangerous one, so that it was not recognized very early as a specific disease. It occurs in a large part of West and Central Africa and also in tropical South and Central America, which it may perhaps have reached in negro slaves. It is estimated that some 20 million people are infected, and a proportion become blind. Blindness rates vary, but in the Volta river basin in West Africa there are thought to be about 70 000 blind adults in a population of 10 million. In some hyperendemic areas up to 20% of adult males are blind. This greatly impairs farming capacity, often reducing to below subsistence level. Villages are deserted and safer areas become overcrowded. The disease is not, however, rapidly contracted, so that travellers and temporary residents in the tropics are rarely affected.

Nature of the disease

As with other filarial infections, patients harbour numbers of long-living adult worms; as many as 100 may be present, living up to fifteen years. In contrast to mosquito-borne filaria, these adults do little harm, though in heavy infections, they form nodules on the body surface. Each female worm produces enormous numbers of micro-filariae, which survive for a year or two. They circulate in the skin and can be seen in tiny snips of skin taken from patients and examined under a microscope. Several hundred may be found in a milligram of skin; they also penetrate to various parts of the eye. Reactions of the patient's body induced by dead micro-filariae cause intense itching, and in due course the elastic layer of the skin is destroyed, giving the wrinkled appearance of old age. The most serious effect, however, is blindness, due to damage of various parts of the eye.

Fig. 5–1 . A blind victim of onchocerciasis in West Africa, being led by a child. The man is not yet forty years old. (Photo by E. Mandelman, courtesy of W.H.O.)

The parasites

The filarial worms responsible for onchocerciasis belong to the species *Onchocerca volvulus*. Similar filarial infections occur in animals, but *O. volvulus* is closely adapted to man and there do not appear to be any animal reservoirs. Micro-filariae are taken up from superficial layers of the skin by the pool-feeding blackflies. The cycle in the insect resembles those of mosquito-borne filariae. A proportion fail to penetrate the stomach wall; successful ones embed in thoracic muscles and eventually migrate to the proboscis, ready to emerge with the next blood meal. This maturation period depends on temperature and can be less than a week in warm conditions. The parasites re-invade man, like mosquito-borne filariae, with the insect's blood feed. They migrate to subcutaneous tissues, undergo further moults and become adult in about a year, but there are no serious effects until repeated infections build up adult numbers (with eventual nodule formation) and prolific micro-filarial production.

The vectors

The blackfly vectors, all species of *Simulium*, are small, sturdily built flies, 2 to 6 mm long (Fig. 5–2). The immature stages, egg to pupa, live in fast-flowing (and thus oxygenated) water. Breeding sites range from small streams to enormous rivers like the Nile or the Congo, usually in stretches of rapids. The adult females take blood meals from various warm-blooded animals. They are especially troublesome near rivers, and hence are liable to infect fishermen and women drawing water or washing clothes. Apart from their importance as disease vectors, blackflies can cause very unpleasant reactions to their bites, and when they are numerous constitute a severe nuisance in parts of the world where no onchocerciasis exists.

Treatment and control

Two drugs are available for treatment of onchocerciasis, but they are not very suitable for large-scale use, so that the most hopeful attack is on the vector. Larvae are destroyed by adding insecticide to head-waters of rivers or streams so that it will pass downstream, killing all blackfly larvae for as much as 150 km. *Simulium* larvae are extremely susceptible to DDT, which has been used with success in many parts of Africa and elsewhere. In Kenya, the local vector, *S. neavei*, was actually eradicated, river by river, and transmission permanently stopped, but the predominant African vector, *S. damnosum*, easily flies great distances from one river to the next. Therefore, if control is to be permanent, a very large-scale operation must be planned to eradicate this blackfly from an entire river basin. At present, such a scheme is in progress, covering the Volta river basin (700 000 km^2) and involving seven West African countries. DDT, however, will be replaced by a less persistent chemical. The work, which is likely to take twenty years, is financed by the U.N. Development Programme and carried out by W.H.O.

5.2 Diseases transmitted by sandflies

Distribution and importance

Sandflies are vectors of three quite different pathogens. Papatasi fever, or sandfly fever, is a viral disease of man, occurring in the Mediterranean region and the Near East; it is of short duration and not dangerous. Carrion's disease, or oroya fever, is due to a bacterium, *Bartonella bacilliformis*, and occurs in the north-western zone of South America, where it probably existed in pre-Columbian days. The fever is severe and often fatal, and after recovery an unpleasant eruption persists for some time, known as verrugia. These diseases, however, are less important than those caused by the flagellate protozoan, *Leishmania*. Different forms of these occur in a wide band all round the tropics and sub-tropics, and these will be discussed in more detail.

Nature of leishmaniases

The pathogens responsible were originally parasites of rodents, in which they cause a mild cutaneous disease. From these hosts, they become adapted to canines and hence, via domestic dogs, spread to man. In the Old World, where it is widely known as Oriental sore, there are two varieties of cutaneous leishmaniasis. One, which is mainly rural and has a reservoir in wild rodents, causes a moist sore; the other, which is urban, causes a dry sore. In South America an unpleasant complication is a tendency to spread to the mouth and nostrils, producing a repulsive deformity known as espundia. This is mainly transmitted from reservoirs in wild rodents to men working in forest areas.

Kala azar is the visceral form of leishmaniasis, which also occurs widely in the tropics and sub-tropics, though seldom in the same areas as the cutaneous form, for some reason. After a long incubation period (sometimes two or three years), irregular periods of fever and various grave symptoms occur, often terminating in death. Fortunately, however, it responds well to treatment. Like the cutaneous disease, kala azar exists in different forms. In India and China, the disease is confined to man, but elsewhere there are animal reservoirs (wild rodents or wild or domestic canines).

The pathogen

The pathogens are species of *Leishmania*, a flagellate protozoan related to the *Trypanosoma*, which will be considered later (p. 39 and p. 46). Certain flagellates of this general type are single-host parasites of insects, and it is presumed that some have evolved into the alternating host forms. The parasitic cycle is as follows. In the vertebrate, the *Leishmania* parasite exists as a rounded cell devoid of a flagellum, but after being taken up with a blood meal by the 'pool-feeding' sandfly, it develops into flagellum-bearing form. In the insect's stomach, the parasites multiply. In primitive forms, reinfection of the vertebrate is by infective fly faeces, but in the human diseases the infective forms move forward and accumulate in the foregut, to be regurgitated when feeding occurs. In both hosts, reproduction is by binary fission and there is no sign of a sexual cycle.

The vectors

The phlebotamine sandflies are small (1.5–2.5 mm) hairy flies, with long slender legs and almond-shaped wings (Fig. 5–2). Both sexes have

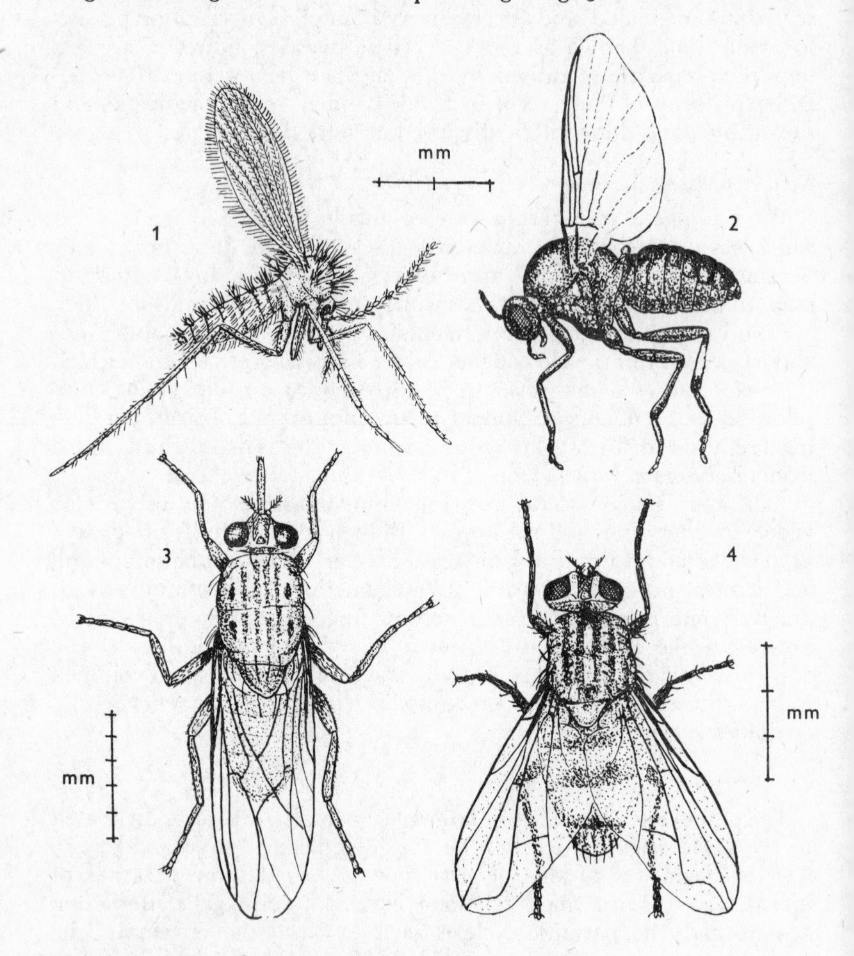

Fig. 5–2 Disease-transmitting Diptera (other than mosquitoes).
1, *Phlebotomus papatasii,* a sandfly; 2, *Simulium damnosum*, a blackfly; 3, *Glossina longipennis*, a tsetse fly; 4, *Musca domestica*, the housefly

rather similar probing mouthparts and feed on plant juices. The females also take blood meals from various vertebrates, to provide protein for egg production. While some prefer birds and reptiles, others favour mammals and sometimes, man, and a few species readily feed in or near dwellings. Sandflies have rather weak powers of flight and probably remain fairly close to their breeding sites. Eggs are laid and the bristly white larvae develop in damp places, among miscellaneous organic debris, especially in crevices among rubble.

Treatment and control

Kala azar is treated by injection of certain antimony compounds. For cutaneous leishmaniasis, antimony treatments are combined with local symptomatic applications to the sores. Immunization against *L. tropica* has been practised successfully in some areas. Against domestic vectors, house spraying with DDT is remarkably successful, since the insect is very susceptible. Often it has been temporarily eradicated as a side result of residual spray treatments for malaria control. Control of reservoirs may involve destruction of infected dogs (which cannot be cured) and eradication of rodents near villages. Forest-living rodents cannot, of course, be eliminated.

5·3 Trypanosomiasis and tsetse flies

History, importance and distribution

Tsetse flies occupy an immense area of tropical Africa, amounting to 4½ million sq. miles, or about one and a half times the size of the U.S.A. Over much of this range, they are the vectors of trypanosomal diseases of man (sleeping sickness) and of horses and cattle (nagana). Although little is known of the Dark Continent before the entry of European powers in the nineteenth century, it is clear that both diseases have had profound effects on human progress in Africa. Sleeping sickness was mentioned in fourteenth-century Arab records, but no idea of the magnitude of its effects was realized until the beginning of this century. Between 1896 and 1906, it is estimated that half a million Africans died of sleeping sickness. Even in the period 1930–46, about half a million cases were recorded in Nigeria alone, out of 6 millions at risk. Despite considerable progress, the disease is still far from being eradicated.

The effects of nagana on man are almost equally drastic, in that through the centuries, Africans have been denied the benefits of domestic animals. Lack of draught animals for ploughing, and of their manure, have handicapped farming; lack of milk and meat has depleted their diet, resulting in widespread protein deficiency.

Nature of the diseases

There are two forms of sleeping sickness due to different species of trypanosome. In Gambian sleeping sickness there are two distinct stages. At first, when trypanosomes proliferate in the blood, there are enlarged glands (see Fig. 5–3), debility and anaemia. After several years the disease usually enters a more severe phase, as the trypanosomes penetrate to the cerebro-spinal fluid. Then there is languor, which gradually deepens, with disinclination to eat, emaciation and finally, coma and death. In the Rhodesian disease, the course of events is much more rapid and death from toxaemia can occur without a long period of

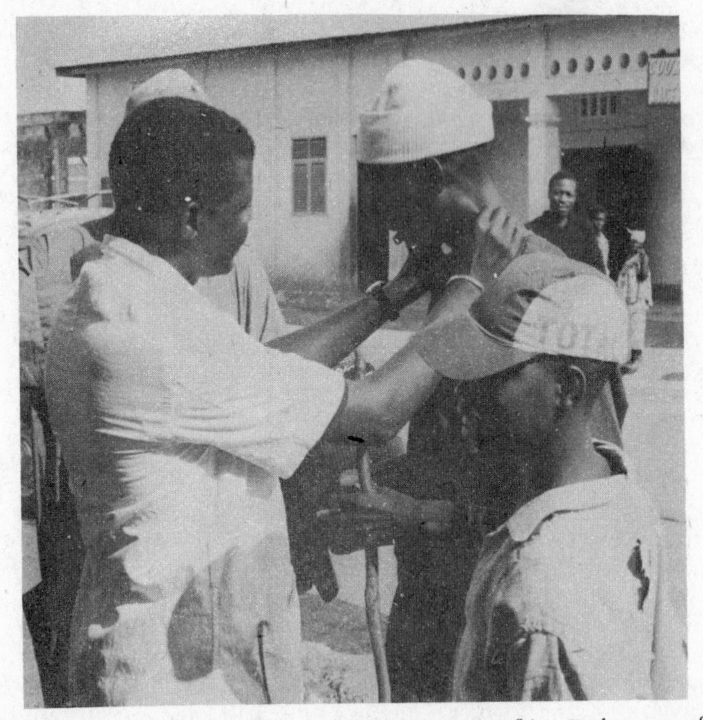

Fig. 5–3 Case detection for sleeping sickness in West Africa. Early stages of the disease can be recognized by enlarged glands in the neck. (Photo by D. Henriad, courtesy of W.H.O.)

nervous degeneration. There is another difference between the two diseases, in that the Gambian form does not involve wild animals, though in some areas domestic pigs may constitute a reservoir. In the Rhodesian disease human cases form a reservoir during epidemics, but an unsatisfactory one because of its acute nature. There is also a reservoir in wild game. As a result, there has been a tendency for men (as hunters) to be infected more than women, and the risk to tourists on safari cannot be disregarded.

The reservoir for the pathogens of nagana is the wild game, especially ungulates, which harbour the trypanosomes without suffering from them.

The pathogens

Trypanosomes are eel-like micro-organisms with an undulant membrane attached along the body, extending forward as a free flagellum. They reproduce by binary fission and there is no evidence of sexual reproduction. Trypanosomes live as blood parasites of vertebrates, with varying degrees of specificity and most are harmless. Nearly all are transmitted by blood-sucking invertebrates, usually 'pool-feeding' biting insects. Those responsible for sleeping sickness and nagana are transmitted by tsetse flies, and infection is by the insect's bite. A rather different type will be encountered later, as the cause of Chagas' disease (p. 45); this is transmitted by the infective faeces of blood-sucking bugs.

Though occasionally transmitted by contaminated mouthparts, most infections of sleeping sickness and nagana involve a definite cycle of development within the tsetse fly before transmission can occur. When the fly becomes infective, it remains so for the rest of its life. Trypanosomes responsible for nagana multiply in the tsetse proboscis or foregut. Those responsible for sleeping sickness undergo a rather complex migration in the fly and end up (after about twenty days) in the salivary glands. Because of the complexity of the cycle, flies infected with sleeping sickness are much rarer than those carrying nagana.

The tsetse flies

Tsetse flies are brownish insects, about the size of a bluebottle, with prominent mouthparts extending forward from the head and visible to the naked eye (see Fig. 3–2 and 5–2). Unlike most other insects, the female does not lay large numbers of eggs, but retains a single fertilized egg in a kind of uterus, feeding the larvae which hatches, by an internal 'milk' gland until it is mature. Then a fully-grown maggot is born and deposited in a suitable shady spot. It immediately burrows into the soil and pupates, finally emerging as the adult fly. Both sexes take meals of vertebrate blood every few days, and no other food. Several species will bite man, though they mainly prefer to feed on wild game. They seek their blood meals mainly by sight and will follow moving vehicles as well as animals and men; also they are attracted to stationary beasts and even to simple lures, such as a blanket hung over a stick.

Various species of tsetse are ecologically adapted to different localities by their choice of feeding terrain and the sites preferred for giving birth. These habitats range from swamp forest to bushy grasslands on the edge of the desert. Very often there are patches of favourable vegetation infested with tsetse, known as 'fly belts'. Cattle-herding tribes know their location and avoid them. Two groups of tsetse species are important for trypanosomiasis. There are some which breed under tress near water and haunt the banks of rivers and lakes. These quite often bite people at fords and watering places and are responsible for Gambian sleeping sickness. Other species breed in open woodland and feed mainly on large game. These are the so-called 'game tsetses', highly dangerous to domestic cattle and also carriers of Rhodesian sleeping sickness.

Treatment and control

Two kinds of drug have been used for trypanosomiasis. One type acts against parasites in the blood and is thus valuable for treating early infections; this type can also be used for prophylaxis. The other type of drug comprises organic arsenicals which, administered carefully, can reach parasites in cerebo-spinal fluid.

Control of sleeping sickness and nagana rests mainly on preventing contact with tsetse flies. A great deal can be done by exploiting the ecological habits of vector species. Thus, several riverine forms can be extensively eradicated by cutting down trees along rivers and lake margins, thus making their breeding sites unacceptable to them. Game-feeding tsetse will not exist in areas from which wild game has been cleared. Hence human settlements and farming must be sharply separated from game reserves, if necessary by fences, and large game must be eliminated from the settled areas. Residual insecticides may be sprayed on fly-infested bush from aircraft, or applied to vegetation by men with knapsack sprayers. Insecticides combined with drug administration are likely to be valuable in epidemics; ecological methods are for more permanent protection from trypanosomiasis.

5·4 Infections spread by non-biting flies

Everyone knows the housefly, the bluebottle, the greenbottle and other blowflies. They are members of a very prolific group of diptera, which frequent dwellings and farm buildings, where man and his domestic animals provide various sources of extra protein for them. Animal waste products form breeding media for various species; some utilize dung, others cadavers, and a few breed in wounds or sores of living animals. The adult flies lick up sweat, lachrymal fluid or blood from small wounds. The housefly invades our kitchen and dining rooms to share our sugar, milk, etc. It can be imagined that these contacts with man and his domestic animals offer opportunities for the transport of pathogens. Yet, for some reason, none of them have developed parasitic cycles in these non-biting flies, to alternate with the higher animals. As vectors, houseflies and blowflies are merely additional means of transmission besides others, such as polluted water or direct contagion. Thus, although various species of flies have been found carrying the germs of poliomyelitis, cholera, leprosy, or tuberculosis, there is no evidence that they are involved in transmission to any extent. Even correlations of fly abundance and disease incidence is not proof, since both could depend, quite separately, on climate. Indeed, the only entirely convincing evidence of the importance of flies as vectors is a decline of disease following fly reduction, which is not seen in untreated 'control' areas. Only two types of disease are definitely implicated in this way; bowel diseases and eye infections, which we must consider separately.

5·5 Bowel diseases

Importance and distribution

The infections in question range from relatively mild attacks of travellers' diarrhoea to potentially dangerous conditions like typhoid, and are due to various pathogens. Generally speaking, they are considerably more common in hot or tropical countries, but even more important is an association with bad sanitation, especially in relation to defaecation. There is a rough analogy to malaria in that new residents in such countries are very vulnerable to infections, whereas the natives are largely immune. This immunity is due to early and repeated infections and is often purchased at the price of infant mortality. Young children are notably sensitive (especially if under-nourished) and there is a Russian proverb to the effect 'Fly in April, dead child in July'.

Nature of the diseases

All these infections are acquired by swallowing infective material and begin their action in the intestines, a common symptom being diarrhoea. In bacillary dysentery, enteritis and some forms of food poisoning, the pathogens remain in the gut, invading and destroying the mucous membrane, causing bleeding and ulceration. In mild cases the result is little more than sickness and discomfort for a few days. More severe infections cause fever, dehydration and, in grave cases, even bowel perforation. Typhoid differs from other diseases contracted in the same way in that the pathogens penetrate the gut and invade blood and other organs. Prolonged fever results, dangerous if untreated.

The pathogens

Most of these diseases are caused by members of the enterobacteriaciae. Several species of *Shigella* are responsible for bacillary dysentery, *S. dysenteriae* being the most virulent. *Salmonella typhi*, which is responsible for typhoid, is restricted to man; but there are numerous other species and strains of *Salmonella* which occur in a very wide range of different animals and can be transmitted from them to man. These can be responsible for various degrees of food poisoning. *Eschericha coli* is a normal comensal inhabitant of the human intestine, but there are virulent strains which can cause dysentery. It seems likely that these are responsible for infantile summer diarrhoea and travellers' diarrhoea. In all these diseases, the pathogens are shed with the faeces. Transmission can occur if drinking water is contaminated, but perhaps more frequently by hands, contaminated after defaecation, when preparing food.

The vectors

Flies very rarely transmit pathogens acquired in the larval stage. Although they will breed in human faeces and the maggots' guts become filled with teeming numbers of bacteria, nearly all are eliminated or destroyed on pupation. It is the habits of the adult female housefly which are important, because she visits both faeces and human food. The simplest mode of transport is via her contaminated feet, but only tiny traces are carried on them and they soon dry up, which often kills the pathogen. Another route is by the so-called vomit drops, due to the fly's habit of partial regurgitation. But perhaps the commonest method of transmission is through the fly's faeces. The passage through the gut is quite rapid and many organisms can pass out unharmed. Defaecation in well-fed flies is frequent, perhaps every five minutes. The process of passing virulent organisms can persist for as long as a week.

Treatment and prevention

Prophylactic immunization may be given for typhoid, which induces antibody formation in the blood, and all these diseases can be successfully treated with various drugs (sulphonamides, antibiotics, etc.). General prophylaxis involves improved hygiene and cleanliness, especially a clean water supply and washing the hands after defaecation. So far as flies are concerned, the best proof of the danger of fly transmission is in reduction of enteric diseases following effective fly control. This was achieved in several different parts of the world in the early days of DDT usage and it was followed by reductions in enteric infections, especially shigellosis. Unfortunately, this can no longer be achieved, due to insecticide-resistance in most houseflies. Ingenious new measures are being considered, but for the foreseeable future we shall have to rely on old-fashioned sanitation. Preventing fly access to faeces is essential. Water closets and good sewage systems are ideal, but much can be achieved by good privy design in backward countries (with deep-bore holes; and screened windows).

5.6 Eye diseases

Importance and distribution

Two types of eye disease are liable to be spread by flies; conjunctivitis, a relatively mild inflammation, and trachoma which may lead to defective vision and even blindness. The history of trachoma is very ancient and it has been said that no eye disease, or indeed no disease of any kind has caused more suffering or personal and economic loss. Like the bowel diseases, incidence of trachoma is characteristic of hot and unsanitary countries and it is particularly prevalent in dry, dusty areas such as North Africa and the Middle East. The W.H.O. estimates that some 500 million people suffer from the disease.

Nature of the diseases

TRACHOMA is an exclusively human disease. In endemic areas, many children are affected. They may recover, but after repeated infections, serious lesions eventually develop. The inside of the upper eyelid is first

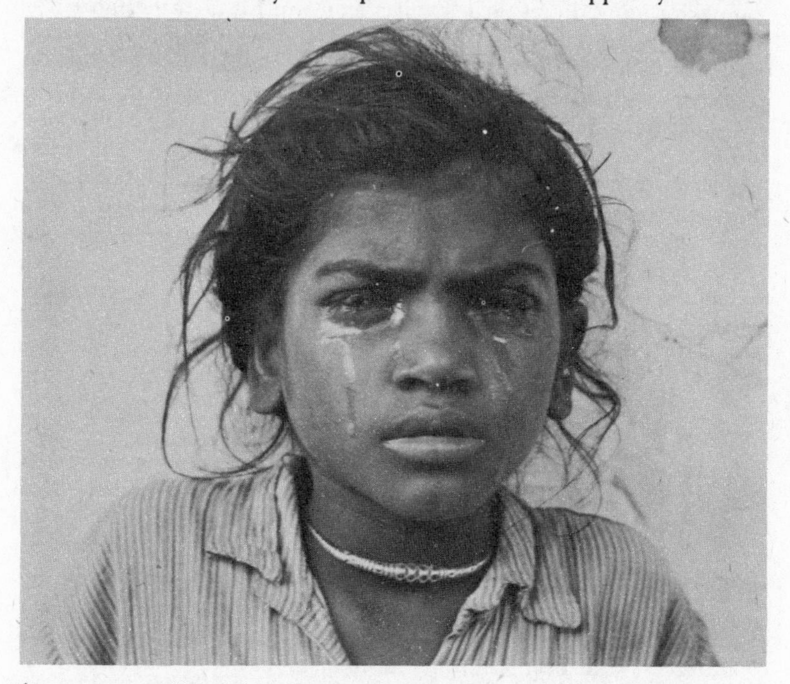

Fig. 5–4 An Indian child suffering from incipient trachoma. (Photo by Homer Page, courtesy of W.H.O.)

affected, but eventually the cornea is involved and covered with granulation and ulceration, so that the sight is impaired. The sources of infection are secretions from affected eyes and purulent discharges from nasal mucous membranes (see Fig. 5–4).

CONJUNCTIVITIS ('sore eyes', 'pink eye') begins with irritation, lacrimation and inflammation of the interior of the eyelids. This is followed by swelling of the lids, pain, avoidance of light and a purulent mucous discharge. Mild infections terminate in a few days, but more severe ones last two to three weeks.

The pathogens

There is some controversy as to the specific identity of the organism primarily responsible for trachoma. It is most probably a virus of the psittacosis-lymphogranuloma group. Conjunctivitis presents a more complex picture, being due to various bacteria.

The vectors

It is most likely that extensive spread of eye disease is due to contagion. But anyone who has witnessed the habits of flies in hot arid countries cannot doubt their involvement (which has furthermore been substantiated by scientific investigation). The faces of young children, in particular, are often covered with flies; they seem more apathetic or indifferent than adults, who usually brush the flies away. Naturally, eyes with suppurating discharge are more attractive to flies than healthy ones and the insects can be seen to feed on the purulent mucus. The species primarily responsible for spread of trachoma are the common housefly and a smaller relative, *Musca sorbens*. The latter breeds readily in human faeces but unlike *M. domestica* it does not enter buildings. In the open, however, its habit of settling on the face is very annoying. Apart from these flies, some distantly related, tiny dark flies are equally troublesome in many warm climates; they are known as 'eye flies'. Though definite proof is lacking, it seems probable that these flies are concerned with the spread of conjunctivitis.

Treatment and prevention

Treatment of trachoma presents problems, since there is no specific drug to counter the initial pathogen, though sulphonamides and antibiotics are useful to counter associated bacterial infection. Surgical treatments may remove some of the visual obstruction. Conjunctivitis being much less grave can often be treated merely by soothing lotion, though severe cases may be helped by suitable bactericides. Prophylaxis against trachoma consists primarily of better washing facilities to ensure clean faces. It has been said that 'an increase in living standards of 1% results in a fall of trachoma incidence of 10%' and 'a water-tap in every village and a bottle of the simplest eye drops in every village would end trachoma in a generation'. This approach would seem more valid than an attempt to control flies (a virtually impossible task in the areas concerned) especially as they are only partially responsible for transmission.

6 Diseases spread by Insects other than Diptera

6.1 Introduction

It is convenient to deal separately with diptera and all other insect disease vectors on grounds of numbers and importance, since the contribution of the two groups is roughly equal. Furthermore, in the diptera the breeding site is relatively remote and the adult insects come seeking their prey, whereas other blood-suckers are more characteristically parasitic, breeding on or near to the host, with a tendency to abandon their wings. The contrast between the mobility of diptera and that of lice and fleas has some influence on the epidemiology of the diseases they transmit. The common bed bug is similarly wingless, but it is not, apparently, a disease vector. Triatomid bugs, however, are intermediate between the diptera and other vectors (Fig. 6–1). They possess functional wings, but do not fly often or

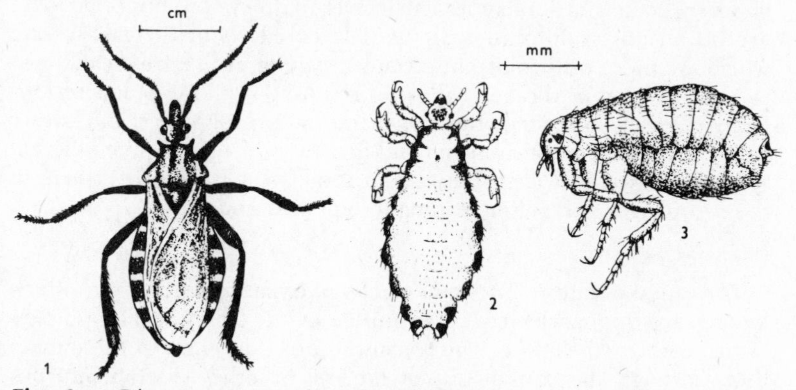

Fig. 6–1 Disease vectors other than Diptera. 1, *Panstrongylus megistus*, a triatomid bug; 2, *Pediculus humanus*, the human louse; 3, *Xenopsylla cheopis*, the tropical rat flea

vigorously and they seem to be on the evolutionary track of the bed bug. Those species which molest man have a similar tendency to infest his dwellings.

6.2 Chagas' disease and triatomid bugs

Importance and distribution

Chagas' disease occurs in South and Central America and is believed to affect about 10 million people. It is chronic and endemic rather than epidemic and acute, so that it is not very dramatic in expression; also it tends to afflict poor, rural peasants. City dwellers and foreign visitors are not at risk, unless they sleep in such bug-infested hovels as Charles Darwin did on an excursion from H.M.S. *Beagle*. (As a result, he is believed to have acquired the disease, which seems to have affected him in later life.) Satisfactory curative drugs are lacking, so that many sufferers are still doomed to early death; and the problem is now receiving considerable attention, not only in South America, but from international bodies such as W.H.O.

Nature of the disease

The pathogen is a trypanosome, which can be transmitted congenitally (from mother to embryo) or, occasionally, in blood transfusions. The usual mode of transmission, however, is by blood-sucking triatomid bugs. The disease can be acquired from triatomids which have fed on other human cases or on infected domestic animals, usually in the same house. But even if all these carriers were cured, new infections could be acquired from a vast reservoir in wild animals (mammals, not birds). In primary cases, which are usually young children, acute and sometimes lethal fever may develop; but often there are no serious symptoms. It is the delayed effects of Chagas' disease which are more disturbing. Infections in young people frequently lead to degeneration of the heart muscles resulting in more or less serious heart disease. Sometimes this degeneration is progressive and causes death in persons between fifteen and fifty years of age. A quite different secondary effect of Chagas' disease in some areas is an enlargement of the oesophagus and colon, leading to digestive troubles.

The pathogen

The causal organism is a small species of trypanosome, *T. cruzi*, which belongs to groups which are transmitted by the faeces of biting insects rather than by their bite. The trypanosomes gain entry to the human body through the conjunctiva of the eye or other external mucous membrane; alternatively, they enter small wounds such as scratch marks (due perhaps to the irritating bites of the insects). In the body, the trypanosomes leave the blood stream and enter muscle tissue, where their lose their flagella and divide by binary fission. They then grow new flagella and re-invade the blood stream. This process of invading tissue to reproduce is repeated periodically. The insects are infected by taking blood containing trypanosomes. A further development cycle takes place in the insects' guts and within about a week or two infective forms occur in the faeces. The bugs remain infective for life, apparently not suffering any harm.

The vectors

The vector status of triatomine bugs for Chagas' disease has some analogy with the anopheline vectors of malaria. A few dozen species have been described in different countries: 33 in Brazil, 16 in Argentina, 13 in Venezuela, etc. Most of them probably are capable of transmitting *T. cruzi*, yet only quite a small number are important vectors, usually one or two in any given region. The main reason (as with malaria vectors) is their invasion of human dwellings. Species which live in trees, feeding on birds or in lairs of wild animals, are unimportant except as agents for maintaining the animal reservoir. There are intermediate species which sometimes invade dwellings; and there are highly domestic forms which usually do so. Unlike the mosquito vectors, however, the triatomine bugs form persistent colonies indoors. They hide in cracks and crevices, which are always numerous in the wattle-and-mud constructed hovels of most rural dwellers in the tropics. At night they emerge and, once or twice a week, wander in search of a blood meal. If, however, they cannot get a feed for any reason, they are very resistant to starvation and can survive for months without feeding, like the common bed bug. Their eggs are laid in harbourages in walls and furnishings and the young nymphs which emerge resemble small copies of their parents, though without wings and sex organs of course. Development is slow and the whole cycle can take anything from six months to two years, according to species and prevailing temperature.

Treatment and control

Since no satisfactory medication is available, much effort has gone into preventive measures, mainly involving house-spraying to destroy the vectors. Several South American countries have instituted campaigns of rural house-spraying with insecticide every year and some have shown positive results in a decline of infection rates. Unfortunately, the emergence of resistance is always a threat, and this has materialized in Venezuela, where the main vector *Rhodnius prolixus* has become resistant to the most effective insecticide (BHC) in several areas. Because of the threat of resistance, there must be some anxiety about the long-term future of control by insecticides. An ideal solution would be substantially improved housing in rural districts, which would render bug infestations easy to see and to eradicate. Some progress has been made in certain countries, for example, Venezuela. But, as with so many tropical problems, inadequate finance and poor educational standards are difficulties to be overcome.

6.3 Bubonic plague and fleas

Importance and distribution

As a result of the horrifying impact of plague epidemics, many historical records are available. A terrible and widespread epidemic occurred about the time of the Emperor Justinian, sixth century A.D., and another in the thirteenth century (the Black Death) which killed a quarter of the population of Europe. Other epidemics, including the 'Great Plague' of London in 1665, flared and smouldered in Europe, gradually dying out in the eighteenth century. The original source of these epidemics seems to have been Central Asia, from which they slowly followed caravan routes. At the end of the nineteenth century, the greatly increased international shipping carried infection from a fulminating epidemic in China, all round the world. At first, sea ports were afflicted, but from them, rural plague foci were established, which remain as persistent threats. With modern methods of treatment and control, however, much of the terror of plague is past history. In the past decade, the total numbers of cases of plague reported to W.H.O. annually ranged between 1000 and 6000, with only 100–200 deaths.

Nature of the disease

Plague is primarily a disease of wild rodents, among which it persists endemically, sometimes flaring into an epizootic which kills many of them. It is transmitted among these rodents by their fleas, but since these are very loath to bite man, the wild rodent disease is not directly serious. (Exceptionally, fur trappers may skin infected animals and acquire the disease by contagion.) Plague epidemics have normally begun by interchange of fleas from wild rodents to domestic rats around the periphery of towns. The disease then spreads rapidly amongst urban rodents, to which it is highly lethal. One of the two common urban rodents, the black rat, is liable to infest wooden houses, though it rarely manages to infest modern dwellings. This brings it close to man, and the last step is a passage to man by species of flea which will bite both rat and man.

In man, plague bacilli sometimes proliferate enormously in the blood, causing septicaemic plague, which is very rapidly lethal. In most cases, however, the bacilli are largely suppressed and localized in characteristic swellings, called 'buboes', usually in the groin or armpit. The numbers in the blood are very low, except for a short period before death in fatal cases. Therefore, man is not infectious to fleas, and bubonic cases do not infect other people. Sometimes, however, a prolific infection in the lungs develops causing the highly dangerous 'pneumonic' form of plague. This causes coughing in the patients and is highly infectious by air-borne droplets, so that inter-human spread occurs without involving fleas.

The pathogens

It is difficult to be sure whether the bacillus responsible for plague, *Yersinia pestis*, is a remote descendant of parasites of mammals or of arthropods. So far as we can judge, the alternating cycle between rodents and their fleas must be very ancient. Yet it has neither evolved into a benign parasitism of the mammal, nor is there a specific development cycle in the insect. What usually happens is a proliferation of bacteria in the flea's gut, forming a gelatinous plug which more or less blocks it. The flea is then unable to swallow its next meal of blood; but as it gets hungrier, it makes repeated efforts to do so. The blood is sucked desperately up to the blocking plug, but usually spurts back into the host animal. This often detaches clumps of bacteria, resulting in a heavy infection of the animal concerned. If blocking of the gut is sustained, the flea invariably dies of starvation, often within a week in tropical conditions. On the other hand, those fleas which do not become blocked, live normally and may get rid of the bacteria. Since, however, the most efficient transmission is lethal to the flea, this suggests an ill-adapted dual parasitism, which is dangerous to both its hosts.

The vectors

The fleas constitute a small and unique order of insects, whose evolutionary affinities are obscure. Though all of them are wingless parasites in the adult stage, they undergo development with complete metamorphosis, indicating a descent from winged insects. The legless grub-like larvae somewhat resemble those of certain diptera. Adult fleas have 'stream-lined' flattened bodies, with backward-directed spurs and bristles, which are ideal for scrambling through the fur or feathers of their hosts (Fig. 6–1). The jumping powers of their muscular hind legs assist both in reaching new hosts and escaping from them if chased. In their choice of hosts for blood meals, different fleas vary in specificity, some being very restricted, others catholic in taste. The so-called human flea, *Pulex irritans*, will feed on and breed near pigs and certain wild animals as well as man. What is important in the present context, however, is that it is very reluctant to feed on rodents, so that it must seldom be responsible for transferring plague from rat to man. Domestic rats are attacked by several species. The common European rat flea seldom feeds on man and is a negligible plague vector. On the contrary, the tropical rat fleas of the genus *Xenopsylla* will very readily bite man and they constitute the main vectors. The danger is especially acute when plague has killed a rat and the infected fleas will readily leave the cadaver, seeking a meal on humans in the vicinity.

Treatment and control

Part of the ancient horror of plague was its highly lethal nature. Mortality ranged from 20 to 90%, depending mainly on constitution and the quality of nursing. With the introduction of sulphonamides and antibiotics, the danger is greatly abated, but it is still a grave disease. Control and eradication of plague involves the surveillance of the residual foci of plague among wild rodents (so-called 'sylvatic' plague) in various parts of the world. It is unlikely that these can ever be eliminated, and it is important to prevent their extension to semi-domestic rodents, and from them to rats in neighbouring towns. In such urban areas, the risk of an epizootic among the rats, with possible spread to man, depends on prevalence of rodents and of *Xenopsylla* fleas on them. Both must be kept under surveillance and controlled so far as possible. Rat control involves elimination by good building construction and rat destruction by poisoning. As regards the latter, the dangerous acute poisons like arsenic have been replaced by safer chronic poisons, like warfarin (though this has been recently hampered by resistance in the rats). When and where plague outbreaks have occurred in recent years, the first measures are directed against the vectors. Liberal dusting of rat runs by DDT and similar insecticides results in contamination of the rats and destruction of their fleas. This usually stops the incidence of human cases and allows the authorities to concentrate on the more lengthy job of rat reduction. Unfortunately, DDT-resistance among *Xenopsylla* fleas has developed in a number of places, requiring the substitution of alternative insecticides.

6.4 Typhus, relapsing fever and human lice

Lice are typical ectoparasites, living permanently on their host. They have become so pampered by living in a comfortable environment, with meals constantly available, that they cannot survive much starvation, nor breed at temperatures much different from their host's skin. Since they are wingless and sluggish crawlers, invasion of new hosts can only occur when the latter come into close contact. These facts basically affect the nature of louse-borne diseases. Living always at skin temperature, they are scarcely affected by climate; indeed, far from benefiting from hot climates or seasons, they proliferate best when cold weather encourages their hosts to wear plentiful clothing and huddle together. The distribution of lice and louse-borne diseases is world-wide; but in tropical countries they are most prevalent at high altitudes. Above all, lice are dependent on low standards of personal hygiene and epidemics of louse-borne disease can only occur in conditions of general lousiness. There are two important louse-borne diseases, typhus and relapsing fever, and the conditions favouring each are substantially similar. Epidemics of the two can be concurrent and relapsing fever was formerly thought to be a variant of typhus.

6.5 Typhus

Importance and distribution

Epidemic typhus has rivalled plague in the numbers of victims claimed in historical times; but its history is different. Circumstantial evidence allows us to recognize some early epidemics of typhus, but there are no definite records of it prior to 1489. Thereafter, however, it seems to have been a constant scourge in a succession of European wars, and in peace time, it fulminated in armies, navies and prisons (as ship fever, jail fever, etc.). The last great epidemic, in eastern Europe following the First World War, is estimated to have affected 30 million people and killed 3 million of them. In recent decades the danger has receded, with a global total of some 10 000 to 20 000 cases a year and a few hundred deaths, mainly in the African continent.

Nature of the disease

After an incubation of about twelve days, a severe fever begins with associated symptoms (headache, nausea, delirium). The patient develops a confused mental condition with a tendency to stupor, and from this state the name is derived from a Greek word *typhos*, meaning hazy. After a few days, a characteristic dull-red, subcutaneous mottling appears on the trunk. The fever lasts about two weeks and the patient either dies of exhaustion or begins a slow recovery. The disease is much less severe in children. Malignance increases with age and malnutrition, with about 50% deaths in middle age and few survivors among old people. Typhus is probably an exclusively human infection. In between epidemics, it may fulminate endemically in remote regions; in addition, the infection can persist in people after recovery and recrudesce when they are under stress. Such long delayed relapses were noted among undernourished immigrants to the U.S.A. about 1900, and were described as Brill's disease. Should they occur in conditions of widespread lousiness, a new epidemic of typhus could be started.

The pathogen

Typhus is caused by *Rickettsia prowazeki*, a name which commemorates two early investigators who contracted the disease and died; the American H. T. Ricketts and the Austrian von Prowazek. When a louse feeds on a person suffering from epidemic typhus, the rickettsiae taken up with the blood meal multiply in the insect's gut and invade the cells of the stomach wall. These cells swell up and burst, releasing large numbers of rickettsiae which pass out in the faeces. This process injures the gut wall and blood perfuses into the body cavity so that the moribund lice turn red. Louse faeces, though damp when emitted, soon dry to a fine black powder. Infected faeces in this condition remain virulent for at least three months. As can be imagined, these faeces become air-borne if infected clothing or bedding is shaken. They will readily cause typhus infection if they are inhaled or enter the eye or a mucous membrane, and another common cause of infection is by invasion of scratch marks or other abrasions. The mode of infection thus resembles that of Chagas' disease, but differs from the latter in causing the death of the vector insect.

The vectors

There are three forms of human lice: the 'head' and 'body' races of *Pediculus humanus* and the 'crab' or 'pubic' louse, *Pthirus pubis*. All forms are able to transmit rickettsiae in the laboratory, but it is the body louse which is mainly responsible for transmitting diseases (Fig. 6–1). The reason for this relates more to habits than to physiological or biochemical differences. Thus, habits of crab lice make them unlikely to spread rapidly through a population. They live on body hair, mainly in the pubic region and are extremely sluggish. A common means of dissemination is in sexual intercourse (the French call them *papillons d'amour!*). Head lice and body lice are very closely related and will interbreed; but their differing habits radically alter their public health importance. Head lice are mainly prevalent on children (especially girls), being disturbingly common in industrialized towns and cities. Body lice live on the under-garment next to the skin and lay their eggs on the seams. They cannot persist on people who regularly launder their underclothes and, in civilized communities, they are restricted to vagrants, of all ages. With the loss of hygienic amenities due to warfare or disaster, body lice can easily spread through the population and create a risk of louse-borne disease.

Treatment and control

Treatment of typhus patients has been very greatly improved by the introduction of antibiotics, but careful nursing and the treatment of adverse symptoms is still very important. Since the discovery that lice were the vectors of typhus (about 1910) it has been realized that the eradication of the parasites is the most effective way of coping with an epidemic. Early methods of louse destruction, by heat or by fumigation, were slow and gave no protection from the rapid reinfestation that usually happened under conditions of widespread lousiness. During the First World War, front line troops were never permanently free from lice, despite the efforts of army hygienists on both sides. In the Naples typhus epidemic of 1943, DDT powder was used for the first time. At numerous disinfestation stations set up by the Allied Forces, thousands of people were treated daily, by the simple process of inserting a dusting gun at openings of the clothing. Not only was this rapid, but the dust in the clothing remained for a week or two, protecting the people from reinfestation by fresh lice. The dramatic success in quenching this epidemic was repeated a decade later in Korea, and elsewhere. Unfortunately, DDT-resistant strains of lice are developing in many parts of the world, so that an alternative insecticide may be needed for future epidemic control. For sanitary personnel coping with louse-borne disease epidemics, special protection is available in the form of immunization by a killed vaccine.

It should be noted that wholesale dusting of citizens is a measure specifically for disease control. It is not suitable for reducing the incidence of head lice among school children, since the obvious signs of a powder application (acceptable in an emergency situation) would be resented under normal conditions. Insecticidal lotions, imperceptible after application, have been developed for this purpose.

6.6 Louse-borne relapsing fever

Importance and distribution

Records of early epidemics of relapsing fever are even more difficult to verify than those of typhus. It was not until about 1840 that the two diseases were recognized as distinct. The causal organisms of relapsing fever were demonstrated about 1870 and those of typhus in 1910. The distribution of relapsing fever is roughly similar to that of typhus; and like that disease, it is being driven back to its last strongholds before, we may hope, eventual eradication. Though a serious complaint, it is much less deadly than typhus, with a mortality usually under 10%.

Nature of the disease

After an incubation of four to eight days, a high temperature develops, with severe headache and other pains. Later a rash of small red spots may be seen, so that the signs and symptoms somewhat resemble those of typhus. Unlike typhus, however, there is a remission of fever after about a week, with sweating, weakness and some relief, but often with evidence of heart strain. There may then follow one to three more relapses into high fever, though these are less dangerous than the first. Louse-borne relapsing fever is exclusively human and in inter-epidemic periods, it continues sporadically in unhygienic areas.

The pathogen

The organism responsible for relapsing fever are parasitic spirochaetes of the genus *Borrelia*. The spirochaetes constitute a diverse group, ranging from free-living forms to parasitic and pathogenic types. The latter, which are distinctly smaller than the free-living species, include forms responsible for syphilis, yaws and pinto. The *Borrelia* group have adopted a double cycle, alternating between arthropods and vertebrates. There are various theories on how this came about. The earliest arthropod vectors seem to have been ticks and the involvement of human lice by *Borrelia recurrentis* seems to have been a later event.

The transmission of louse-borne relapsing fever is as follows. Spirochaetes in the blood of a febrile patient are sucked up by the feeding lice. Many of them are digested, but a proportion of them survive and begin to appear in the body cavity of the louse (between gut and cuticle) after about six days. They remain in the body cavity of the louse for the rest of its life, apparently doing no harm to it. They cannot escape and get transferred to another human host until the louse's skin is broken, allowing the infectious fluid to escape. This may happen when infested people destroy lice by bursting them between the fingernails, or even crushing them between the teeth, a habit often observed among primitive peoples! Infection occurs by penetration through mucous membranes or through scratches, as in the case of typhus. But one important difference will be noted, there is no danger of infection from the louse's faeces, since these are not infective.

Vectors, treatment and control

The observations on typhus vectors and their control are all relevant to louse-borne relapsing fever. Immunization is not practised.

7 Diseases transmitted by Ticks and Mites

7.1 Types of ticks and blood-sucking mites

Disease vectors among the acarina mainly occur in the sub-order Ixodoidea or ticks, all of which feed on the blood of vertebrates. There are two kinds of tick, the argasidae, or soft ticks, and the ixodidae, or hard ticks; they differ radically in life history and habits as well as in morphology (Fig. 7–1). In the soft ticks, the mouthparts and bases of the legs are completely hidden by the oval sack-like body, the appearance of which makes them resemble greyish, leathery buttons. The life cycle and

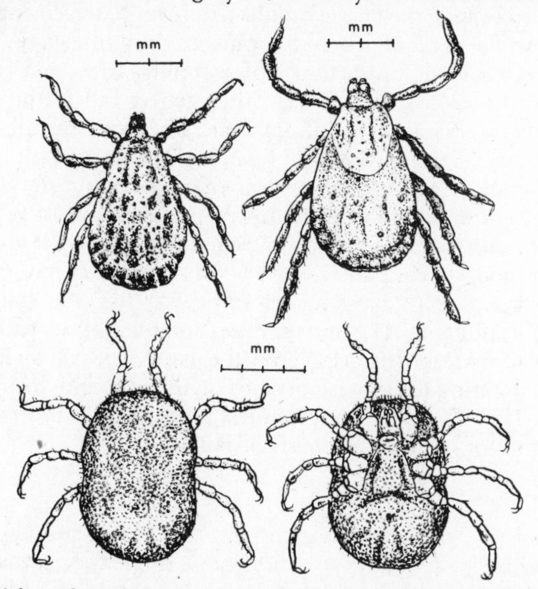

Fig. 7–1 Ticks. **Above**, *Dermacentor andersoni*, a hard tick (left, male; right, female). **Below**, *Ornithodorus moubata*, a soft tick (left, dorsal; right, ventral)

habits have some aspects in common with blood-sucking bugs. Large colonies tend to develop in proximity to the sleeping place of the host and during the day they usually hide in crevices. Eggs are laid in batches and the three-legged larvae emerge after anything from a week to a month. Following the larval stage there are four or five nymphal stages with four pairs of legs like the adult. All stages (except non-feeding larvae in some species) have similar feeding habits. They emerge at intervals to take blood meals, but if this is denied, they are extraordinarily resistant to starvation and some species can survive five to seven years after a single meal. They are not only tolerant of starvation but can endure very arid conditions and are also difficult to kill by chemical pesticides.

Hard ticks are distinguished not only by their projecting mouthparts, but by the possession of a hard, shiny shield, which covers nearly the whole back of the male, but only part of that of the female. These ticks spend much longer periods on the host animal, gorging with blood. The females drop to the ground indiscriminately, so that large areas of open country may be infested. They lay enormous quantities of eggs (3000 to 8000 are common) and then die. The larvae from these eggs eventually climb up herbage, waiting for a passing host to crawl up. Many fail to encounter a host and die, but the vast egg numbers compensate for this. In some genera, the rest of the life cycle is spent on the same host ('one-host ticks'), in others, the larvae moults to a nymph, which engorges and then drops to the ground, where it moults to the adult stage. The adults then have to await the opportunity of grappling with, and feeding on, another host ('two-host ticks'). Yet other genera fall to the ground to moult between larva and nymph, as well as between nymph and adult ('three-host ticks'). Hard ticks are less resistant to starvation than soft ticks and, in most cases, less tolerant of arid conditions. It must be remembered, however, that in the micro-environment of vegetation at soil level, conditions can be much more humid than in the ambient air.

Various blood-sucking mites, such as the red poultry mite (a common pest in chicken roosts) have habits somewhat like the soft ticks and colonize nests and lairs. The harvest mites and their allies, with which we will be concerned shortly, resemble the hard ticks in their habit (as larvae) of remaining for longish periods on the host and dropping down at random. They therefore tend to infest areas of country frequented by the hosts and not merely their nests or lairs.

7.2 Acarines as vectors

Transmission of micro-organisms by blood-sucking acarines (especially ticks) is rarely very specific, since a single species may transmit various pathogens and several pathogens are each spread by different species. Virtually none of these blood-sucking acarines is solely a human parasite and the diseases which they transmit are, in most cases, maintained in animal reservoirs and only sporadically transmitted to man. A wide range of types of pathogen is involved. Some of those most dangerous to man are allied to forms transmitted by insects, notably, the louse-borne spirochaetes and rickettsiae and the mosquito-borne arboviruses. Other types of pathogen include bacteria (*Francisella tularensis*, causing tularaemia) and protozoa (*Babesia* and *Theileria* responsible for cattle diseases) and filarial parasites of small animals transmitted by ticks and mites.

Most of the organisms responsible for serious disease in man or domestic animals are apparently harmless to their acarine vectors. Many of these vectors are remarkable for their ability to pass the infection from the larval stage through to the adult and often, through the egg, to the next generation. In ticks, these characteristics are usually combined with a long life, so that they can constitute an important reservoir of disease. Several acarines feed very slowly, thus extending the opportunity to become infected. Further transmission from ticks may be by several routes; to vertebrate hosts by persistent salivation, by extrusion of coxal fluid, or copious defaecation; to other ticks by trans-ovarial infection or via infective semen during copulation.

7·3 Ticks and relapsing fever

Distribution and importance

Tick-borne relapsing fever can occur within a wide band round the tropics and sub-tropics. In Central and South-East Africa, there may be local human reservoirs of the disease, but elsewhere it is primarily a disease of rodents and transmitted only incidentally to man. The human disease is transmitted by soft ticks which infest the earth floors of primitive dwellings. The animal disease is transmitted by ticks which infest the lairs of rodents or other small mammals, which may be in open country or in caves. In contrast to the louse-borne disease, tick-borne relapsing fever is sporadic and endemic. Travellers, hunters and campers who acquire the disease from animals, or from sleeping in infested huts, are liable to suffer severely as it is a dangerous disease. Among Africans, the indigenous human disease tends to be less malignant.

Nature of the disease

The symptoms of the disease resemble those of the louse-borne form. After a short incubation period, the primary attack begins with severe headache and the usual concomitants of high fever. After four or five days (or less with the African form) the temperature falls. Subsequent relapses may occur after short intervals or after as long as three weeks. In untreated patients, there are usually three to six relapses, though there may be as many as eleven in the African form.

The pathogens

A considerable number of species of *Borrelia* has been mentioned in relation to tick-borne relapsing fever; the endemic African form is known as *B. duttoni*. When the *Borrelia* are taken up in a blood meal by a tick, they gradually invade many organs of its body. Of considerable importance is the infection of the salivary glands which results in transmission during a subsequent feed. Another possible route of infection is via the coxal fluid, which is emitted during or soon after feeding in several species of tick. Infection thus differs radically from that in louse-borne relapsing fever. The soft ticks concerned may live several years and can remain infective for life, so that they constitute a reservoir of infection, apart from the various animals and birds infected with *Borrelia*.

The vectors

Soft ticks of the genus *Ornithodorus* are vectors of tick-borne relapsing fever. Many species are mainly involved with transmission among animals and rarely infect man. The most frequent human infections are due to *O. moubata*, which is common in native huts in a large part of South and East Africa (Fig. 7–1).

Treatment and control

Treatment of patients suffering from tick-borne relapsing fever is similar to that for the louse-borne disease. Control of the endemic African infection consists in thorough application of an acaricide to walls and floors of infested dwellings. Not many chemicals are capable of destroying these rugged ticks; *gamma* BHC is one of the most effective, safe substances. For protection against accidental infection from 'wild' ticks, people camping in infested areas should sleep on camp beds raised from the ground. Luggage, clothing and equipment should be carefully inspected to make sure that they do not harbour ticks after visiting suspected localities.

7·4 Ticks, spotted fever and tick typhus

Distribution and importance

A group of related rickettsial diseases are transmitted by ticks. They include Rocky Mountain spotted fever in North America, *fievre boutonneuse* of the Mediterranean basin, and various kinds of tick-typhus in other parts of the world. All of them are essentially zoonoses, human infections being accidents due to bites of infective ticks and are not transmitted further. Cases are therefore sporadic, and not very numerous. Over a seven-year period (1947–54) there were about 4700 cases in North America, while less than 500 cases were reported from South America during this period. None of the diseases are particularly serious, except the Rocky Mountain fever, which is distinctly dangerous if untreated, with a mortality rate of about 20%.

Nature of the diseases

The reservoirs of Rocky Mountain fever include a considerable variety of wild animals, which do not suffer severely. In addition to the animal reservoirs, the ticks harbour infection through their long life cycle without harm and they can transmit it to a subsequent generation through the egg. Human cases are mainly in rural areas among campers, hunters and others liable to tick-bites; sometimes, however, infected adult ticks are carried by dogs into suburban gardens. After an incubation period of three to seven days, an attack of fever begins with associated chills, headache and eventually a typhus-like stupor. A rash of rose-coloured spots appears at the fourth to seventh day over most of the body. In severe cases, mainly in older people, death may occur with symptoms of heart failure.

The *fievre boutonneuse* seems to have a main reservoir in dogs. This and the other varieties of tick typhus are much less malignant than the Rocky Mountain fever and are not fatal. Another difference is that a small primary sore appears at the site of the tick bite, which may turn gangrenous.

The pathogen

The pathogen responsible for Rocky Mountain fever is *Rickettsia rickettsiae*, while *fievre boutonneuse* and other types of tick typhus are ascribed to *R. conori*, sometimes considered a sub-species of *R. rickettsiae*. The rickettsiae taken up with a blood meal by a tick disseminate themselves throughout its tissues without apparently harming the creature. Invasion of the salivary glands ensures the possibility of further animal infections and penetration of egg cells in the ovaries allows transmission to the progeny. Rickettsiae may be emitted with faeces, but these mostly lose virulence rapidly. After a meal of infected blood, there is a period of about twelve days during which the rickettsiae multiply and subsequently the ticks are infective indefinitely.

The vectors

The vectors of tick-borne rickettsial disease are all hard ticks. Two- or three-host ticks are more liable to encounter and transmit infection than a single-host tick. Forms that cause human disease are likely to be abundant species and must be likely to bite man as well as the natural reservoir hosts. Different species fill these requirements in various parts of the world. The main vector of Rocky Mountain spotted fever in the far western U.S.A. is the wood tick *Dermacentor andersoni* (see Fig. 7–1).

Treatment and control

Treatment of spotted fever or tick typhus resembles that of louse-borne typhus and generally involves broad-spectrum antibiotics. Prevention simply consists in avoidance of tick bites. For people specially at risk, protective clothing, preferably treated with tick repellents, may be advisable.

7·5 Mites and scrub typhus

Distribution and importance

Scrub typhus occurs over a large part of South-East Asia, from India to Japan and south through the Malay peninsula as far as Northern Australia. In Japan it is known as *tsutsugamushi* or 'mite disease', the association with mite bites having long been suspected among indigenous peasants. Prior to the Second World War, it was considered a localized disease, affecting mainly agricultural workers. The war brought large numbers of highly susceptible foreign soldiers into affected areas, as a result there were many alarming and apparently inexplicable outbreaks of what was then a very dangerous disease. The disease is still endemic in many parts of this vast range and it is difficult to see how it can ever be eradicated. Cases are only sporadic, however, and modern treatments have removed most of the danger.

Nature of the disease

The infection is maintained among small rural and forest animals, being transmitted by bites of mites of the family trombiculidae. From these wild creatures, which are not seriously affected, the infection can spread to rodents which inhabit areas frequented by man. These include not only comensal rats, but also field mice and voles. Human infections are acquired in such rodent-infested areas, which often consist of 'scrub' land between recently cleared agricultural land and jungle. After an infected mite bite, there is an incubation period of four to ten days. The site of each infective bite develops into a small ulcer. This bears no relation to the severity of the disease, which varies from mild to very grave.

The pathogen

The causal organism, *Rickettsia orientalis*, is distinct from those of the typhus group and the spotted fever pathogens. The rickettsiae are taken up by larval mites and invade all their tissues. They pass through the nymphal and adult stages and are transmitted through the egg to the larvae of the next generation. This is essential, since the mites only feed once on vertebrates in each generation. In the larvae, the rickettsiae invade salivary glands and infection is transmitted with the saliva.

The vectors

Scrub typhus is transmitted to man by larval mites of the genus *Trombicula* (see Fig. 7–2). These mites are related to the harvest mites of Europe (also described as *Herbstmilbe, Bêtes rouges* and 'Chiggers') which cause intensely irritating bites. The adults are relatively large, velvety mites, orange or red in colour and entirely harmless to man. They lay eggs among soil debris and from these emerge the six-legged larvae which cause the trouble. The larvae crawl up vegetation and seize any opportunity to attach themselves to passing animals or men, crawl up the legs and embed their mouthparts in a suitable tender area. On man, they usually collect along constrictions of the clothing, such as a belt or garter. The larvae feed, not on blood, but on lymph and partly digested tissue fluids. After a period of several days they detach themselves and drop to the ground. Here they moult to the nymphal stage which, like the adult, lives harmlessly upon tiny soil creatures. Over a dozen species of *Trombicula* attack man, and in the study of scrub typhus some difficulty was experienced in connecting the easily collected parasitic larval forms with the harmless, free-living adult. It seems that two closely related species are the vectors of this disease; *T. deliensis* and *T. akamushi*, the former being most widely distributed.

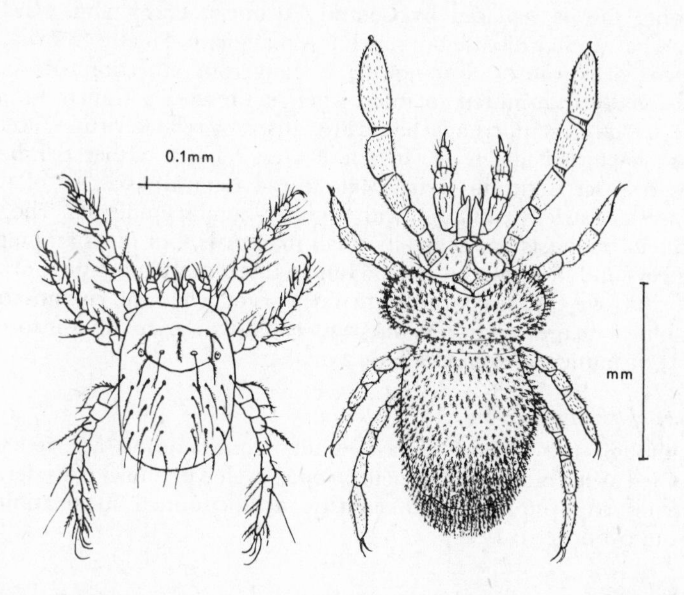

Fig. 7–2 *Trombicula akamushi*, a mite vector of scrub typhus. **Left**, the larva; **right**, the adult

Treatment and control

Satisfactory treatment is now available by the use of antibiotics, though if they are administered too early, they prevent the acquisition of immunity to the disease. Control of the mite vectors may be desirable where camp sites or new buildings are to be put in areas known to be infested. Local danger spots are often recently cleared areas on the edge of the forest. All secondary vegetation and bush should be cut and burnt and the ground treated with mite killing compounds (e.g. BHC). Personal protection for individuals working in such areas may be obtained by treating clothing with miticides and repellents.

7.6 Ticks and virus diseases

As mentioned earlier (p. 27), ticks are second only to mosquitoes as vectors of arboviruses; thus, of some seventy or eighty arboviruses known to affect man, ticks are vectors of about a dozen. These are responsible for a variety of diseases varying in severity and including both haemorrhagic and neurotropic types. Most of the important ones occur in the Old World. Throughout the vast evergreen forests of the U.S.S.R. they are represented by Russian spring-summer encephalitis (RSSE), a dangerous disease which can cause 25–30% mortality. In Europe, this is replaced by Central European encephalitis (CEE) a somewhat similar disease but much less malignant. Finally, in Britain, a related virus causes louping ill, a dangerous infection of sheep, occasionally transmitted to man, when it presents a similar form to CEE. Certain haemorrhagic fevers are caused by related viruses, notably Omsk haemorrhagic fever, in South-West Siberia. Other tick-borne virus diseases occur in India, Malaya and Central Asia. All of these diseases are strictly zoonoses, with no inter-human epidemics. The wild maintenance hosts are generally small mammals, but in Asian jungles, monkeys and birds are involved. Human infections tend to be sporadic, but explosive outbreaks can sometimes occur when the circumstances combine a heavy infection in ticks with extensive agricultural or forestry work, or numerous camping holidaymakers.

Nature of the diseases

Although the various diseases under consideration are generally classified as haemorrhagic or neurotropic, each type shows considerable variation in symptoms and severity, as mentioned in relation to mosquito-borne viruses (p. 27).

The pathogens

Several important tick-borne viruses belong to serological Group B, while others are assigned to the Miscellaneous Group category (see Table 1, p. 8). The serological classification, however, though it may indicate some evolutionary relationship, gives no indication of the type of disease produced. Virus is taken up by ticks feeding on hosts during periods of viraemia, and their prolonged feeding habits increase the chances of this. It persists for long periods in various tissues, being able to over-winter at low levels and proliferate after feeding periods in the spring. Trans-ovarial transmission occurs sporadically and has been established only for some viruses. Infection of vertebrates is generally through the saliva of the tick during biting. Much virus is excreted in the faeces but the importance of this is not known. The virus of European tick fever can be transmitted in milk of infected goats and this represents an additional hazard to rural dwellers.

The vectors

Although soft ticks can transmit some arboviruses, the important vectors are all hard ticks (*Dermacentor, Hyalomma*, etc.) The spring-summer fever group are spread by species of *Ixodes*; these ticks are familiar to many country people in Britain, since *I. ricinus* (the sheep tick) sometimes attaches itself to dogs, or even people, in sheep-farming country.

Treatment and control

Treatment is not specific; good nursing and the relief of symptoms are all important. Special medical care is needed in cases of shock following haemorrhagic fevers. Vector control is extremely difficult with tick-borne arboviruses, though some trials of DDT applications to infested areas have been made in Russia and Central Europe. Some immunizing vaccines are available and some protection for tick bites can be obtained by smearing tick repellents on outer clothing.

8 Vector Control: Achievements and Setbacks

The overall benefits achieved by new insecticides in the first fifteen years after the Second World War drew an encomium from Professor B. G. Maegraith at the VI International Congress of Tropical Medicine and Malaria (Lisbon, 1959). 'Modern insecticides', he said, 'have revolutionized preventive medicine in the tropics.'

These outstanding achievements date mainly from the first two post-war decades. In more recent years, two drawbacks have become evident; the emergence of resistant strains of vectors and the concern about environmental contamination. Mass media repeatedly dwell on the pollution of the environment by synthetic chemicals, an idea with instant appeal to the lay mind, combining as it does, the instinctive love of Nature with an uneasy distrust of Science. Public concern, though sometimes exaggerated, has had the beneficial result of stimulating the formation of official bodies in many countries (as well as sections of the international agencies, W.H.O. and F.A.O.) to scrutinize all possible hazards from pesticide usage. An immense amount of data has been collected and published as reports, apart from a considerable number of books.

8.1 Environmental residues

Only a very few points can be mentioned here; they nearly all concern the chlorinated insecticides which have proved so efficient and been so widely used. The complaints are not on grounds of demonstrable toxicity to man (DDT is one of the safest insecticides known in this respect), but because of the discovery of minute traces, widely dispersed in the environment. This discovery was due to the development of gas-liquid chromatography, an astonishingly sensitive technique. It has revealed traces of DDT in air and rainwater, albeit at a level of 12^{-12} (parts per million million). Substantially greater quantities are found in soil and even in tissues of living animals, especially the fat which accumulates such residues. The amounts in human fat are usually a few parts per million, which is almost certainly of no significance, since people who have worked in DDT manufacture for fifteen years have residues fifty times as great without showing any signs of harm. Indeed, no official report has found evidence of harm to human health from this source; though there are undoubtedly numerous incidents of poisoning of wild life by insecticides. On these grounds, as well as because of a lingering unease about human body residues, it is clearly right for developed countries in temperate regions to reduce the use of organochlorine insecticides as far as possible. They have virtually no arthropod-borne disease, they are not desperately short of food, and they can pay for more expensive alternative insecticides. For tropical countries with big disease problems, food shortage and modest resources, this policy is unsound. DDT has already saved some 10 million lives, with no definitely proved harm to man. It can scarcely be abandoned until alternatives are available. As a final point, it should be noted that agricultural uses constitute by far the greater source of environmental contamination by insecticides. In vector control, the chemicals are used in much smaller quantities and, in many cases, applied inside dwellings, rather than being broadcast over fields.

8.2 Insecticide-resistance

The earliest records of insect pests developing resistance to a pesticide date from about 1910, but until after the Second World War, not more than about half a dozen examples were known. With the introduction of new synthetic insecticides, cases began to increase rapidly until now about 200 species of pests are involved, half of them being of public health importance. These include thirty-five anopheline mosquitoes, eighteen being important malaria vectors, and about the same number of culicines (including *Aedes aegypti* and *Culex fatigans*). Among other important vectors may be mentioned lice, fleas (including *Xenopsylla cheopis*), triatomid bugs and houseflies. The reason for this great acceleration was undoubtedly the exceptional selection pressure, due to the very wide use of the new insecticides, and their residual action. One may note that these factors are also responsible for their persistence in the environment and in both cases, the organochlorine insecticides are mainly involved. Unfortunately, it is the very properties which allow for wide usage and long residual action, which have enabled these chemicals to achieve their results, so that the same disadvantages are likely to characterize any other effective pesticide.

A few years ago (1969) the author collaborated with a W.H.O. scientist in reviewing the current resistance situation in the world, as it affected vector-borne disease. It was concluded that much of the problem resulted from the heavy dependence in previous decades on organochlorine insecticides, notably DDT, BHC and dieldrin. As a result, resistance to these convenient insecticides had developed to the point where there was a severe hindrance to control, in many areas, of the vectors of malaria, yellow fever, filariasis, typhus and plague. Certain other diseases (onchocerciasis, Chagas' disease and sleeping sickness) were not then affected by resistance but signs of resistance in vectors of Chagas' disease and onchocerciasis have since appeared. As a result of wide resistance to organochlorines, more and more use of organophosphates and carbamates was being made. But resistance to these too is growing.

Counter measures against resistance in disease vectors have been co-ordinated by W.H.O. The primary need for accurate and comprehensive information necessitated the use of uniform methods of detecting and measuring resistance in all countries. Suitable techniques were designed for a wide range of vector species and W.H.O. supplied test kits to field workers involved in vector control. The widespread use of such tests has provided a valuable picture of the status of resistance throughout the world. One hopeful sign has been the fact that resistance often does not extend throughout the range of a given species, even when insecticide has been used throughout the area. This suggests that the potentiality for developing resistance may not be universally present in all insect populations.

Despite much brilliant research, no practical remedy for any case of resistance has been devised. The only solution is to change to a different type of insecticide, but the alternatives are limited. The W.H.O. has supported a ten-year search for new types of contact poison, but no outstandingly different compounds were found. Insecticide-resistance clearly calls for intensive research of several kinds, for unless it is overcome, it will gradually drive us back to less efficient control measures. Naturally, considerable thought has been given to ways of using insecticides with care and discrimination at the right time and place, so as to combine adequate control with minimum selection for resistance. It is an unfortunate fact that the more wholesale agricultural uses of pesticides have often accidentally involved insects of medical importance and thereby induced resistance (e.g. by contamination of mosquito breeding places).

8.3 Alternatives to control by pesticides

The growing problem of insecticide-resistance, together with general concern about environmental contamination, have stimulated scientists into a consideration of a wide range of more or less ingenious alternatives. The possibilities are too complex and diverse to discuss in any detail, but a few words may be said about some of them.

1. CONTROL BY PARASITES OR PREDATORS This concept was already well established when the author was a student forty years ago, so that it has been thoroughly explored. The main difficulty is that where the parasites or predators already exist, they have already achieved a balance with the pest and are unlikely to exterminate their source of survival.

2. PATHOGENIC MICRO-ORGANISMS represent a similar means of biological control, but sometimes disease organisms can have drastic effects on pests (e.g. the myxomatosis virus on rabbits). This subject is

now being actively explored, but clearly demands caution, since some pathogens could conceivably mutate and affect beneficial insects, or even vertebrates.

3. INSECT DEVELOPMENT INHIBITORS Synthetic mimics of natural insect hormones can be used to upset the normal processes of moulting or metamorphosis, and certain other compounds disturb the insect's physiology at these critical times. Because of their specific effects on insects, they are virtually non-toxic to vertebrates. Some of these chemicals are now marketed for practical use; but contrary to expectation, insects are able to develop resistance to them.

4. THE RELEASE OF STERILIZED MALES to mate with wild female insects and prevent their reproduction was a remarkable American innovation. To compete successfully with wild males, enormous numbers have to be reared and sterilized by radiation. Unfortunately, owing to formidable technical difficulties, the method has only succeeded in one case, though the idea was conceived over thirty years ago and has been tried repeatedly. Rather similar difficulties apply to releasing males of an incompatible strain of the pest which mate with the wild females but produce infertile offspring.

5. GENETICAL MANIPULATIONS involving induced mutation and selection, seeks to find ways of introducing deleterious genes into pest populations. Alternatively, if a vector can be displaced by a strain of the same species which does not transmit disease, the result would be a triumph.

Unfortunately, most of these new lines of research have encountered intractable problems in practical trials. It seems most unlikely that, at least in the next decade, any of them will challenge the benefits to health achieved by insecticides. One reason for this is that, whereas the same insecticide treatment can be used against a number of different pests, the new methods are highly specific and each one requires intensive research to determine the feasibility of each proposal. Therefore, it seems that pesticides will have to be relied on for vector control for some years, though new methods will gradually supplement them.

Further Reading

ARTHUR, D. R. (1962). *Ticks and Disease.* Pergamon Press, Oxford.

BAKER, J. R. (1973). *Parasitic Protozoa.* Hutchinson, London.

BUSVINE, J. R. (1966). *Insects and Hygiene.* Methuen, London.

CAMERON, J. W. M. (1956). *Parasites and Parasitism.* Methuen, London.

GILLETT, J. D. (1971). *Mosquitoes.* Weidenfeld & Nicolson, London.

GORDON, R. M. and LAVOIPIERRE, M. M. J. (1962). *Entomology for Students of Medicine.* Blackwell, Oxford.

HIRST, L. F. (1953). *The Conquest of Plague.* Clarendon Press, Oxford.

MATTINGLY, P. F. (1969). *The Biology of Mosquito-borne Disease.* George Allen & Unwin, London.

MUIRHEAD-THOMSON, R. C. (1968). *Ecology of Insect Vector Populations.* Academic Press, London and New York.

NASH, T. A. M. (1969). *Africa's Bane: the Tsetse Fly.* Collins, London.

WIGGLESWORTH, V. B. W. (1964). *The Life of Insects.* Mentor Books, London.